GAME THEORY AND CANADIAN POLITICS

This is the first book-length application of game theory to Canadian politics. It uses a series of case studies to illustrate fundamental concepts of game theory such as two-person and n-person games, the Nash equilibrium, zero-sum and variable-sum games, the paradox of voting, the Condorcet winner, the Condorcet extension, the Banzhaf power index, and spatial models of competition. No mathematics more complex than simple algebra is required to follow the exposition.

The case studies are not just contrived illustrations of abstract models but intensively researched studies of important episodes in Canadian politics. Topics include the Lubicon Lake land claim stalemate, the formation of national political coalitions, the adoption of the metric system, nomination of party leaders, the importance of procedural rules in Parliament, and the entry of the Reform Party into the political system. In each case, utilization of game-theory models produces new and often surprising conclusions.

Game theory, and the rational-choice paradigm of which it is a part, are an increasingly important addition to the conventional modes of political analysis. This book is intended to show what game theory can add to the philosophical, institutional, and behavioural approaches that have dominated previous works on Canadian politics.

THOMAS FLANAGAN is Professor of Political Science, University of Calgary. He is the author of several books, including *Louis 'David' Riel: 'Prophet of the New World.'*

THOMAS FLANAGAN

Game Theory
and Canadian Politics

UNIVERSITY OF TORONTO PRESS
Toronto Buffalo London

© University of Toronto Press Incorporated 1998
Toronto Buffalo London
Printed in Canada

ISBN 0-8020-4094-2 (cloth)
ISBN 0-8020-7946-6 (paper)

Printed on acid-free paper

Canadian Cataloguing in Publication Data

Flanagan, Thomas, 1944–
  Game theory and Canadian politics

  Includes bibliographical references and index.
  ISBN 0-8020-4094-2 (bound)   ISBN 0-8020-7946-6 (pbk.)

  1. Political science – Methodology.   2. Game theory.
  3. Canada – Politics and government.   I. Title.

  JA72.5.F53 1998   320.010971   C98-931661-0

An earlier version of part of Chapter 3 appeared in Ken Coates, ed., *Aboriginal Land Claims in Canada: A Regional Perspective* (Toronto: Copp Clark Pitman, 1992). Used by permission Addison Wesley Longman Publishers. An earlier version of part of Chapter 6 appeared as 'Amending the Canadian Constitution: A Mathematical Analysis,' *Constitutional Forum* 7, 2–3 (1996): 97–101. Used by permission. An earlier version of part of Chapter 8 appeared as 'The Staying Power of the Legislative Status Quo: Collective Choice in Canada's Parliament after *Morgentaler*,' *Canadian Journal of Political Science* 30, 1 (March 1997): 31–53. Used by permission.

University of Toronto Press acknowledges the financial assistance to its publishing program of the Canada Council for the Arts and the Ontario Arts Council.

Contents

PREFACE   vii

1 Rational Choice   3

2 Game Theory   20

3 Stalemate at Lubicon Lake   37

4 Models of Metrication   55

5 How Many Are Too Many? The Size of Coalitions   74

6 Who's Got the Power? Amending the Canadian Constitution   93

7 The 'Right Stuff': Choosing a Party Leader   105

8 The Staying Power of the Status Quo   120

9 Invasion from the Right: The Reform Party in the 1993 Election   140

10 What Have We Learned?   164

NOTES   173

INDEX   187

Preface

The invention and elaboration of game theory is one of the great intellectual achievements of the second half of the twentieth century. Although there was some crucial earlier work, the founding event was the publication in 1944 of *The Theory of Games and Economic Behavior*, by John von Neumann and Oskar Morgenstern.[1] Von Neumann was one of the mathematical giants of our century, for whom game theory was an intellectual diversion; Morgenstern was an economist with an interest in mathematics.

The half century following their original book saw a profusion of literature on game theory, particularly in mathematics and economics, but increasingly also in other disciplines, such as political science, psychology, sociology, and biology. The maturity of game theory was recognized in 1994 by the award of the Nobel prize in economics to three eminent scholars in the field. In fact, the award would have come years earlier except that one of the three, John Nash, whose creative contributions came in the early 1950s, fell prey to schizophrenia late in that decade and did not recover until the mid-1980s. An award that did not include Nash was unthinkable, and the Nobel committee did not want to make the award until they were certain that Nash was in a proper state to receive it.[2]

Economists played a pioneering role in applying game theory to the traditional concerns of political science. The first game theory article in a political science journal, published in the *American Political Science Review* in 1954, was co-authored by a mathematician and an economist.[3] William Riker, already a well-known authority on American government, read that article and became the first political scientist to promote game theory within the discipline.[4] For 40 years thereafter, Riker, his students, and an ever-growing number of colleagues have worked to bring game theory into political science.

A simple statistic illustrates their success. In 1995, the *American Political Science Review* (*APSR*), generally considered the discipline's most prestigious journal, published 50 articles. Of those 50, 16, or 32 per cent, were based wholly or partially on game theory.[5] Of course, this statistic from one journal does not describe the discipline as a whole; the percentage of game-theory articles would undoubtedly be lower in most journals, in the United States as well as in other countries. But to the extent that the *APSR* publishes trend-setting articles, it shows the direction in which the discipline may be moving.

In Canada, however, the influence of game theory has not been nearly as great. The 1995 issues of the *Canadian Journal of Political Science* contained 22 articles, of which only one (4.5 per cent) had any connection to game theory, and that was rather remote. We are still in the very early stages of the application of game theory to the study of Canadian politics.

This book is written in full awareness of being virtually at the beginning. Many books on game theory claim to be simple or nontechnical, requiring only a clear mind and high-school algebra; yet I know from reading such books and trying to teach them to intelligent students that their promise of simplicity is seldom fulfilled.[6] This book, in contrast to existing works, is meant to be understood by political scientists and political science students who have no special gift for mathematics. It could be assigned as supplementary reading in a course in Canadian politics or as one of the texts in an elementary course in game theory. I don't claim that it will teach game theory in a systematic way, but I believe it will impart some of the flavour and show what game theory can do. If it inspires some students to go further, it will have achieved its purpose.

I hope that *Game Theory and Canadian Politics* will be read not only by students taking courses but also by others with a serious interest in Canadian politics, for the case studies assembled here present interesting and nonobvious results. These case studies are not just potted illustrations of how to apply techniques; they incorporate considerable research and are meant to convey new knowledge. The use of game-theory models provides a framework for insights that might arise from intuition but that, without such a framework, would remain scattered and fragmented rather than becoming part of a systematic body of knowledge.

Chapters 1 and 2 present the basic ideas of rational choice and game theory. Chapter 3 develops a two-person ordinal game in which the Nash equilibrium is the appropriate solution concept. Chapter 4 uses Schelling curves to make two-person models applicable to society at large. Then come five chapters dealing with topics in n-person game theory. Chapter 5 discusses coalition theory, including William Riker's size principle. Chapter 6 presents the Banzhaf power index as a way of measuring power in minimum winning coalitions. Chapters 7 and 8 both deal with the phenomenon of cycles in n-person voting games. And chapter 9 uses spatial models as a way of representing games with continuous choices. Finally, Chapter 10 reviews the findings of the case studies to show how they complement the philosophical, institutional, and behavioural approaches that dominate the contemporary study of Canadian politics.

The case studies are admittedly eclectic: the Lubicon Lake land claim stalemate; adoption of the metric system; national political coalitions; the distribution of power under various constitutional amending formulas; the nomination of party leaders; the importance of procedural rules in Parliament; and the Reform Party's political breakthrough in the 1993 election. The topics were chosen because each one lends itself to analysis through a different game-theory model. Taken together, they illustrate some of the most important concepts of elementary game theory: zero-sum and variable-sum games, the Nash equilibrium, two-person and n-person games, cooperative and noncooperative games, the Banzhaf Power Index, the paradox of voting, the Condorcet winner, the Condorcet extension, and spatial models of competition. Anyone who works through the variety of illustrations will start to gain some appreciation of what game theory can (and cannot) do for the study of Canadian politics.

I would like to thank Dan Liebman for his careful copy-editing, especially for his painstaking work on the many tables and figures in the manuscript. I am also grateful for various contributions from other staff members of the University of Toronto Press; their professionalism continues to uphold the Press's reputation as the leading academic publisher in Canada.

GAME THEORY AND CANADIAN POLITICS

1

Rational Choice

Game theory belongs to a family of methodologies variously known as rational choice, public choice, social choice, and collective choice. At bottom, all are spinoffs from the discipline of economics, employing the concept of *homo economicus* or 'economic man,' first developed by John Stuart Mill. Political economy, according to Mill, studies man 'solely as a being who desires to possess wealth, and who is capable of judging of the comparative efficacy of means for obtaining that end.'[1] Rational choice and related approaches attempt to use this simplified concept of human nature to explain a much wider range of behaviour than economists usually study.

Economic man is an abstraction, a set of assumptions meant to capture important aspects of human behaviour. Rational-choice analysis consists of teasing out the logical implications of these assumptions, treating them as explanations of how human beings behave in certain situations. The assumptions, reasoning process, and explanations generate models of behaviour, which can be tested against empirical data. No one expects such a model's predictions to be perfectly accurate, because the assumptions are understood to be simplifications.

Economic man is not real man, as Mill knew perfectly well; human beings act under a 'variety of desires and aversions' in addition to those focusing on economic efficiency.[2] But, Mill argued, the way to deal with this variety is to isolate a motivation analytically, to determine what would happen if such a motivation operated by itself. At a later stage, other factors can be brought back in to produce a more realistic model. Although some practitioners have made more exaggerated claims, rational-choice analysis does not have to maintain that human nature is exhausted by the concept of *homo economicus*. Rather, it asks this question: to the extent that we are rational, self-interested utility maximizers, how can we be expected to behave?

Rational-choice models provide a simplified guidebook to political reality, showing how people would act politically if they were motivated exclusively by the pursuit of self-interest. The rational-choice guidebook is of some value almost everywhere, because self-interest is always present in human behaviour. Yet the guidebook never offers a complete description, because human behaviour is driven not only by self-interest but also by altruism towards relatives and friends, attachment to community, and normative obligations arising from customs, traditions, religious beliefs, and laws.[3]

Rational-choice analysis can model any situation, but the model may be of little empirical value if the element of self-interest is small. One of the present, and perhaps permanent, limitations of rational choice is that it lacks within itself a reliable way of discriminating between situations where it is useful and where it is not. Thus rational choice has generated some wildly inaccurate predictions about real-world behaviour. Perhaps the most famous is Anthony Downs's argument that the rational voter would not bother to vote, because the personal costs in terms of time and effort outweigh any conceivable individual benefit.[4] Needless to say, people continue to vote. Similarly, just before the explosion of new social movements in the late 1960s, Mancur Olson made a highly plausible rational-choice argument that such groups should be impossible to organize.[5]

These two predictions failed because the behaviour they tried to model is significantly affected by a combination of altruism and normative obligation. Yet the models are useful even in failure. By giving us a clear picture of how a purely self-interested actor would behave, they point us toward the other factors that affect the outcome in these situations.

For those wanting a full and highly readable introduction to rational choice in political science, the most recent account is Shepsle and Bonchek, *Analyzing Politics*.[6] In this chapter, I merely draw out some basic ideas required to understand the applications to Canadian politics made in the rest of the book.

Methodological Individualism

In focusing on choices, rational-choice methodology assumes that individual human beings are the true agents of politics, because only individuals can make choices. When we speak of collective choices, as made by a parliament or by the voters in an election, we are really speaking of an aggregation of individual choices; that is, counting yeas and nays under a specified decision rule.

Of course, institutions exist – political parties, interest groups, bureaucracies, and all the rest. But rational choice does not take such institutions for granted; it prefers to ask how they emerged from the choices of individuals. Moreover, it is cautious about regarding institutions as actors in their own right. Institutions can be considered rational actors only if some identifiable person, such as the leader of a party or the president of a corporation, is able to make decisions that others in the institution must adhere to. In particular, rational choice is sceptical about speaking of vague aggregates such as 'society' or 'the nation,' because such terms are often used to disguise very real differences among the people who make up the collectivity. 'Canadian society' or the 'Canadian nation' cannot decide anything, because no one is charge. From a rational-choice point of view, a statement such as 'Canadian society supports public health care' means that many Canadians, but not necessarily all, agree with legislation passed by Parliament (a team of individuals led by the prime minister and cabinet) at a certain time. Bringing in 'society' is only a manner of speaking and does not add any information.

Within the rational-choice research paradigm, game theory most often plays the role of providing *microfoundations*. That is, game theory is used to create a decision-making model of a simplified situation thought to capture the essential features of the larger problem under analysis. Institutions are understood as part of the context, creating both positive and negative incentives for individual action. Game theorists do not see their simplified models as ultimate explanations but as starting points for analysis and aids to clear thinking.

Ordinal Utility

Rational choice thinks of human beings as decision-makers facing an endless series of choices between alternatives. Each decision-maker has what is known in the jargon of economics as a *utility function*. A function in mathematics is a relationship between variables, as in the equation $y = x + 2$, which states that for any value of variable x, variable y will be 2 units larger. A utility function is a statement of the relationship between the alternatives in the choice set and the preferences in the mind of the decision-maker.

The simplest kind of utility function is purely ordinal; that is, it only produces a ranking of alternatives. As a voter, you can say that you prefer the Liberals to the Reform Party without saying that you evaluate the Liberals at 10 and Reform at 5, or attaching any other measure of value to the alternatives. An ordinal utility function is a mapping of alternatives in the choice set into a preference ordering in the mind of the decision-maker. Three assumptions about such ordinal decision-making are absolutely essential to any rational-choice analysis: comparability, exhaustiveness, and transitivity.

Comparability means that the decision-maker is able to choose between any two options, even if the choice is one of indifference. Adopting some simple mathematical notation, let us say that = means 'is indifferent between,' > means 'is preferred to,' and < means 'is less preferred than.' Then the axiom of comparability means that, for any options a and b, a decision-maker will be able to say $a = b$, $a > b$, or $a < b$.[7]

Suppose you are voting in a federal election in an electoral district in Atlantic Canada where the three main candidates are from the Liberal, Progressive Conservative (PC), and New Democratic (NDP) parties. The axiom of comparability means that you are able to decide whether you are indifferent among the three or, for example, prefer the Liberal to the PC candidate, the PC to the NDP candidate, or the NDP to the Liberal candidate. Note that all these are decisions; you do not say, 'I can't make up my mind.' In simplest terms, comparability means you can always make up your mind among options, even if making up your mind leads you to decide you are indifferent.

Exhaustiveness is an extension of comparability. It implies that decision-makers can produce a complete or exhaustive preference ordering if there are more than two relevant options in a choice. To go back to the election example, exhaustiveness means that you will be able to produce a statement such as, 'The NDP candidate is my first choice, the Liberal is my second, and the PC is my third,' or 'I am indifferent between the Liberal and the PC candidate, but I would prefer either one to the NDP,' or something similar. In other words, you can deal with all the options and sort them out into a hierarchy or preference ordering.

The axiom of *transitivity* implies that, if $a > b$, and $b > c$, then $a > c$. That is, if you prefer the Liberal candidate to the PC, and the PC to the NDP, you also prefer the Liberal to the NDP. If the PC candidate withdraws from the race, you now have a choice between Liberal and NDP. If you then vote Liberal, you show that your preference structure is transitive: Liberal > PC > NDP. If you vote NDP, your preference structure is intransitive because you are reversing your previously established preference of Liberal over NDP.

In fact, decision-makers in the real world do experience many preference reversals over the course of time. Comparability, exhaustiveness, and transitivity apply only to decision-making in specific situations within which it is reasonable to assume stability of preference structure. In the case of voting, that stable situation might be a single election, or it might be durable over periods of years or decades.

Comparability, exhaustiveness, and transitivity are the essential logical properties of *rationality*, as that term is understood in economics and by extension in the rational-choice approach to politics. Economic man is rational man in the sense that his decision-making has these three properties. Note, however, that these are all formal properties of how we make choices, not substantive characteristics of what we choose. Some writers call the former concept 'thin rationality.'[8]

We can imagine a thinly rational person who thinks that the whole world is controlled by aliens from another planet, and that the earthlings who appear to occupy positions of power are merely puppets. Such a person might formulate the following preference ordering: assassinate the Pope > assassinate the prime minister > commit suicide. Most of us would find this preference ordering bizarre and irrational in the everyday sense, but it could be thinly rational in terms of embodying comparability, exhaustiveness, and transitivity.

Emphasizing the formal and ignoring the substantive properties of rationality allows us to factor out the divergent preferences of real human beings. We commit ourselves to looking for the common properties of all decision-making, whatever the objects of desire might be. Exponents of rational-choice methodology believe this emphasis leads to important generalizations about human behaviour in all kinds of diverse circumstances.

Cardinal Utility

Economists sometimes talk about 'utiles' as units of utility in the minds of decision-makers, but no objective measuring scale of this kind exists. However, for many purposes, we can reasonably assume that preferences will be proportionate to some measurable quantity, so we can use such measurements as a proxy for utility. For an investor wishing to maximize rate of return, the ratio of profit to dollars invested is an excellent guide to making decisions. Other things being equal, we would expect such an investor to prefer a bond paying 8 per cent interest to one paying 7.5 per cent, and so on. We could thus have a utility function based on a cardinal (i.e., numerical) measure, rate of return. Or a military commander fighting a war of attrition might use a body count of opponents' casualties as a measure of success, in which case tactics producing a higher body count would be preferred to ones producing a lower count.

The basic assumptions of comparability, exhaustiveness, and transitivity all apply to cardinal utility functions, but they do not suffice; a number of other assumptions also have to be brought into play. For our purposes, these can be simplified into the statement that rational actors with cardinal utility functions employ the probabilistic logic of expected value. To introduce some simple notation, let V be the value of some outcome to the decision-maker, P the probability that this outcome will occur, and EV the expected value. Then the expected value equals the product of the value and the probability with which it occurs, or $EV = P \times V$.

All of this is much easier to grasp with an illustration. Suppose you are given a choice between two alternatives: A is a guaranteed payoff of $1000, and B is a 50 per cent chance of winning $2000 in a lottery. Rational choice assumes that you are indifferent between these outcomes ($A = B$) because the expected value of both is $1000 ($1000 x 1.0 = $1000 = $2000 x 0.5).

Assuming that rational actors reason in terms of expected value allows us to import the immensely powerful apparatus of probability theory, facilitating all sorts of mathematical analysis. The only problem with the assumption is that we know it is not always true! To vary the earlier example, suppose that you are offered a choice between a guaranteed payoff of A = $1000 or B = a 1 per cent chance of winning $110,000. Under this assumption, you should prefer B to A, because the expected value of B is $110,000 x 0.01 = $1100, which is greater than the expected value of A = $1000 x 1.0 = $1000. Which would you prefer? If you are a typical student in need of money, you probably would take the sure thing. On the other hand, gamblers seeking excitement and a big payoff have been known to wager large sums of money on worse odds than one in 100.

A lot of laboratory research has been devoted to sorting out the circumstances under which people do or do not use the logic of expected value, and some alternative approaches are now available.[9] However, they are mathematically complex, and no generally applicable rival to the standard approach has emerged. At the introductory level at which this book is written, models of cardinal utility functions normally use the logic of expected value in a straightforward way.

Self-interest

In market transactions – the chief domain of economics – an assumption of self-interest does not create many problems. It is obvious that in general sellers prefer higher revenues, workers prefer higher wages, customers prefer lower prices, and so on. People enter the market precisely to serve their self-interest. Individual exceptions may exist but do not affect the overall functioning of the market. There is the complication of collective economic units such as the family or the firm, in which several or many people pool their efforts, but this factor doesn't change things fundamentally. Singles tennis is an individual sport, while basketball is a team sport; but players in both sports are equally motivated to win, even though winning in basketball requires cooperative activity. The same sort of reasoning can be applied to collective units in the sphere of economic competition.

There is also an economics of altruism. Remember Mother Teresa in the slums of Calcutta? In ministering to the poor of India, she was scarcely pursuing her self-interest in any normal sense of the term; she lived her life altruistically, for the sake of others. Nonetheless, she must have had to make hard decisions. Which beggar got a bowl of soup? Which invalid got to take up a scarce bed in the clinic, and which was sent home with a little medicine? In making such decisions, Mother Teresa undoubtedly used both ordinal and cardinal utility functions, depending on circumstances. She was surely aware of the price of medicines and would not have used them where they would have had no effect. She must have learned to sort out the deserving poor from unscrupulous pretenders. She could thus be rational in all her choices even though she was not pursuing self-interest in the usual sense.

Things get even more complicated in the study of politics because those who engage in political action almost always profess a variety of altruistic motives.[10] They claim to be working on behalf of a particular group such as farmers or native peoples, or on behalf of the whole community, or in pursuit of abstract principles such as freedom, justice, and equality. You never hear politicians openly say they are seeking office because they enjoy power, because they need the salary and pension, or because they want to give patronage jobs to their relatives and friends.

There is, therefore, a real problem in defining the payoffs in models of political behaviour. Do we take the altruistic statements at face value, or do we assume that they are merely a cover for the true, self-interested motives? In one of the pioneering works of rational choice, *An Economic Theory of Democracy,* the economist Anthony Downs took the second path. He made the explicit assumption that, regardless of what they say, politicians in a democracy are motivated by the desire to get elected and re-elected, because that is how they can 'attain the income, power, and prestige which come from being in office.'[11] Politicians say they seek office in order to implement certain policies, but Downs reversed the relationship: '[P]arties formulate policies in order to win elections, rather than win elections in order to formulate policies.'[12] Downs knew perfectly well, and explicitly admitted, that this was not the whole truth about politicians; it was a simplifying assumption made to construct a consistent model of political behaviour.

Similarly, Brennan and Buchanan argue that rational self-interest is the only feasible assumption to make in interpreting political behaviour. 'What do individuals seek in politics,' they ask, 'if they do not seek to maximize their own expected net worth?'[13] Anyone who proposes a different working assumption would have to show that it applies to a broad spectrum of political behaviour, not just to isolated exceptions. Most rational-choice work on politics proceeds on this basis. It postulates a hard-boiled goal for political actors – getting elected, maximizing votes, getting special treatment for interest groups in legislation, winning court cases, raising money, and so on – and then estimates the extent to which this self-interested goal explains actual behaviour. This book follows the general practice of assuming 'thick,' or substantive, rationality, based on self-interest, in addition to 'thin,' or formal, rationality, based merely on consistency.

A related point is that actors' utility functions depend only on their own evaluations. One can write $y = f(x)$ and not $y = f(x, z)$, where x is the payoff to the actor and z is the payoff to someone else. The political actor's goal may incorporate payoffs to many others, but he pursues it as his goal in competition against other actors with different goals. He may be forced to compromise and take less than he is aiming for, but that is not because he wants others to win; it is because that is all he can get under the circumstances.

Some authors have tried to envision a methodology of rational-choice altruism in which actors take the goals of others into account and utility functions have the form $y = f(x, z)$, but this approach leads to much more complex models.¹⁴ In this book, I will assume that politics is a contest of many players, all of whom are striving to win, however they may define winning.

Perfect Information

At the introductory level of rational-choice analysis, we assume that actors have all the information relevant to the situation. They not only know what they want to achieve but they also know what their competitors want, and each side knows that the other knows. There are no secrets; it is like chess, where the whole gameboard is visible. Moreover, everyone involved knows the relevant connections between cause and effect; that is, if I do this and you do that, certain things will happen.

The notion of perfect information is obviously an oversimplification, and much of the more advanced work in rational choice consists of relaxing this assumption and devising more sophisticated models to deal with imperfect information. You can think of the use of simple probability as a step in that direction. For example, assume that major-league baseball managers see batting average as the appropriate measure of success for their hitters. In 1995, the Toronto Blue Jays' first baseman, the left-handed-hitting John Olerud, batted .304 against right-handed pitching but only .259 against left-handers.¹⁵ Because Olerud's chances were demonstrably poorer against left-handers, the Blue Jays' manager often chose to play someone else at first base when the other team started a left-handed pitcher. Similarly, the opposing team would sometimes bring in a left-handed pitcher against Olerud in a difficult situation. Managers on either side could not predict whether Olerud would get a hit in any particular at-bat; but, knowing the long-run probabilities, they could act so as to improve the outcome for their side. In everyday language, they could 'play the odds.' This book will incorporate a few examples of playing the odds into some relatively simple models, but will not go deeply into the complexities of risk and uncertainty, which is a task for a more advanced and systematic study of rational choice.

Let me close this section with an example that occurred to me a few years ago while spending a month at the University of Michigan. I was living in a subleased apartment and taking the bus back and forth to campus each day. At the local bus stop, I had a choice of two routes – call them *A* and *B* – that would take me to my destination. Route *A* took

23 minutes for the trip, route B only 15. But the buses on the slower route A offered better service. They ran four times an hour, specifically at 8:03 AM and every 15 minutes thereafter. The buses on route B ran only three times an hour – at 8:07 and every 20 minutes thereafter. Which bus should I take?

To answer this question, I had to create a model of the situation. First, I decided that my utility function would depend solely on the total travel time – time waiting for the bus plus time sitting on the bus. The principle of transitivity implied that I would always prefer shorter time to longer time. At that point, I only had to work out the decision rules that would specify when $A > B$, $A < B$, and $A = B$.

It was not difficult to compute total travel time. Suppose, for example, that I arrived at the bus stop at 8:17. I would have to wait one minute for bus A, which was due at 8:18, and then travel 23 minutes on the bus, for a total time of 24 minutes. In comparison, I would have to wait six minutes for bus B, which was due at 8:27, and then travel 15 minutes, for a total travel time of 21 minutes. The rational choice was clear: I should let bus A go, wait for bus B, and save three minutes overall.

The model could be interpreted in two ways, either normatively or predictively. Normatively, it told me what I *should* do if I wanted to use my time as efficiently as possible. Predictively, it told me how I *would* behave in the real world of bus-stop behaviour. Here is where things started to get interesting. Despite understanding the normative aspect of the model perfectly well, I found that I had a strong tendency to get on the first bus that pulled up, even if my mental calculations projected a longer total travel time than I would get by waiting for the next bus.

That my behaviour diverged from the normative ideal suggested that my simple model overlooked some significant aspect of the situation. What was ignored, I think, was that buses are not always on time. When a bus pulled up, I could be reasonably sure how long it would take from that point on; but I was less certain of how long the trip would take if I waited for the next bus, because that one might be late arriving. In theory, I could have adjusted for this factor in my model by adding an average lateness term in calculating the total travel time for the second bus. To go back to the original example, I should get on bus A at 8:18 if the average lateness of bus B was more than three minutes, because then the total *expected* travel time via B would be more than the quite certain 24 minutes via A.

This choice amounts to using the logic of expected value as described above. However, I had no data on the lateness of Ann Arbor buses, and a one-month stay in the city did not seem long enough to allow me to compile my own observations. In the absence of good information to justify a probabilistic solution, I fell back on the folk wisdom expressed in the proverb, 'a bird in the hand is worth two in the bush.' Take the first bus that shows up; who knows when the next one will come? To use the term introduced by Herbert Simon, I was *satisficing* rather than optimizing.[16] Rather than take the trouble to look for the absolutely best solution, I was opting for a course of action shown by experience to be good enough for the circumstances.

This homely example illustrates a couple of important limitations of rational-choice analysis. First, an initial model is often too simple to deal with the complexity of the actual situation, and that defect shows up in inaccurate predictions about real-world behaviour. Second, a more realistic model is often difficult to construct, either because the necessary information is missing or because complicated and costly calculations would be required. Under such constraints, it can be rational to satisfice by relying on habits, customs, or traditions that have worked more or less well in the past in a variety of analogous situations.

I also chose this example for what it is not. Although it illustrates the rational-choice approach to decision theory, it is not an example of a game because there was no opponent. Neither the bus company nor the bus drivers were trying to outwit me (even if it sometimes seemed that way). I did not have to take anyone else's decisions into account when making my own. I was in full control of my destiny, at least on the average. This type of rational choice is sometimes called a game against nature, but it is not a game in the usual sense of the term. In a true game, the outcome is not dependent on the decisions of one player but is *interdependent* on the decisions of two or more players. Games are the subject of the next chapter; but before going on to that topic, we have to take a closer look at what rational choice can contribute to the study of politics.

Value Added?

Rational choice is a relative newcomer to the study of politics. Although some of the pioneering works by economists date from the 1950s, the whole enterprise was peripheral to political science when I was a student in the 1960s. It is now fashionable in the United States but still marginal in Canada. As with all new methodologies, it is wise to be cautious. Can rational choice teach us things about Canadian politics that we can't learn from approaches that have been in use much longer – indeed, for centuries or even millennia?

The case studies in the rest of this book illustrate why I think the answer is yes. But one can also try to answer the question in a more general way. Let us look briefly at three of the most common approaches to the study of Canadian politics, each having its own strengths and weaknesses, and see how rational choice can complement them.

Philosophical

These approaches are the philosophical, the institutional, and the behavioural. Because political philosophy has such a long and rich tradition, extending back to Plato and Aristotle, some political scientists devote themselves almost exclusively to the study of these writers, reading their works carefully and comparing them with one another. Such research has much in common with history, philosophy, and literary criticism. It leads to a deeper understanding of texts; but, if it stops there, it does not tell us much about Canadian politics, or indeed about any politics. Note, however, that the great political philosophers wrote chiefly about politics, not about other philosophers. We still study Aristotle because he was a keen observer of politics, not because he was a trenchant critic of Plato (though he was that, too). In our own day, the leading practitioners of political philosophy also write about politics, not just about other authors.

An excellent Canadian example is Charles Taylor, professor of philosophy at McGill University. Taylor is a world-renowned authority on the philosophy of Hegel, but he has also written extensively on Canadian politics. To simplify a great deal, Taylor has concluded philosophically that modern politics is largely about identity, both individual and collective. Assertion of identity, however, cannot be meaningful without recognition from others. Moreover, there are many different political identities in a country like Canada, leading to what Taylor calls deep diversity.

All of this, in his view, has practical implications for the continued existence of Canada. Quebec's political leaders may demand power and money from the rest of Canada, but what they really want, he says, even if they are not consciously aware of it, is recognition.[17] The best way to preserve Canada would be to put a distinct society clause in the Constitution as a symbolic recognition of Quebec's uniqueness, and then to practise asymmetrical federalism, allowing Quebec to be different and to develop those differences further.

The point here is not whether Taylor is right or wrong, but the nature of the enterprise he is engaged in as a political philosopher. The hallmark of the activity is abstract reflective analysis. Although Taylor knows a great deal about the history and politics of Canada and Quebec, and that information forms an indispensable backdrop to his thinking, he does not marshal it systematically to test empirical hypotheses. Also, the analysis ends with a kind of moral exhortation or normative imperative. For Taylor, recognition of deep diversity is not just a means of preserving the Canadian state, although he sometimes presents it that way; it is obviously the way he thinks politics should be conducted.

Abstract reflective analysis and moral exhortation are typical of attempts to study politics philosophically; they are the strengths, but also the weaknesses, of the method. Abstractness leads to intellectual clarity, and moral exhortation sets standards for how we ought to act; but the clarity may be deceptive, and the standards may not correspond to how people actually do behave. Taylor makes a powerful case for why Canadians should embrace the symbolism of distinct society, but he has less to say about why that concept is so unpopular outside Quebec, or why it is also dismissed by Quebec nationalists themselves. At times he seems to take refuge in saying, 'They just don't get it' (more elegantly expressed, of course).

Institutional

Much political science consists of writing painstaking descriptions of institutions such as cabinets, legislatures, political parties, interest groups, and the mass media. These are the collective actors on the political stage, and no one can understand politics without knowing a great deal about them. At a low level, this sort of work can be almost purely descriptive, detailing the key positions in organizations with their powers and responsibilities, or writing the biographies of major figures within the institutions. At a higher level, the researcher can analyse the logic of the institution, showing how organizational incentives lead individuals to act in certain ways even when they may wish to do otherwise.

To take a Canadian example, C.E.S. Franks, in *The Parliament of Canada*, explains why party discipline remains so strong in the House of Commons, even though the major parties have all proposed, at one time or another, to relax it.[18] Regardless of what politicians say, the institutions of responsible government, which require the cabinet to maintain a working majority in the House, steer politicians toward acting cohesively. Advancement within the party depends on being a good team player, on pursuing goals defined by the party leadership. Members who cannot come to terms with this practice will vegetate on the back benches, be expelled from caucus, or retire in disappointment.

Institutionalism has produced some of the most important findings of political science, but it also has limitations. As an example, practitioners may devote so much time to studying the Supreme Court of Canada that they may lose sight of how different other courts may be. In theory, at least, this problem can be overcome by maintaining a comparative perspective. More fundamentally, institutionalism needs to be complemented by data on how individuals behave, and by a theory of how institutions arise out of individual actions. Behaviouralism, described below, can fill the first gap; rational choice can fill the second.

Behavioural

The so-called behavioural revolution of the 1950s and 1960s imported into political science a large toolbox of techniques previously developed in psychology and sociology: sample surveys, attitude scaling, content analysis, inferential statistics, and much more. Although there was some resistance at the time, political science would be unthinkable today without the behavioural approach, which underpins extensive knowledge of public opinion, political culture, and electoral behaviour.

Behavioural research typically takes the individual human being as the focus and unit of analysis, and proceeds to look for correlations between behaviours and attitudes, demographic characteristics, or other behaviours. Much of the early research on voting behaviour was devoted to establishing correlations between partisan loyalty and demographic characteristics such as sex, race, religion, income, or education.

With the aid of advanced statistical techniques of multivariate analysis, behavioural research can become highly sophisticated in its ability to control for extraneous factors; but without an adequate theory of human nature it remains basically correlational. And as all statistics teachers tell their students, correlation is not causation. Much behavioural research in political science is rather like the first epidemiological studies of AIDS in the early 1980s, which found the famous '4H' risk groups: homosexuals, heroin users, hemophiliacs, and Haitians. The nature of the first three groups suggested that transfer of body fluids might be involved, but the risk factors did not really make sense until HIV was discovered and an etiology for AIDS could be proposed.

In political science, it is now easy to compile an immense amount of information about the correlates of voting, but it is much harder to explain why the correlations sometimes change dramatically and on short notice. Harold Clarke and colleagues, whose decades of survey research constitute a notable achievement of Canadian political science, resort to the concept of 'permanent dealignment' to describe Canadian voters in the 1990s – an honest admission of inability to predict what will happen next.[19] Their attempt to link this phenomenon to systemic factors, by arguing that decades of brokerage politics have made voters cynical, is not entirely satisfying. If permanent dealignment is really caused by brokerage politics, why did it not happen earlier, and why not also in other countries?

Another problem with behaviouralism, not surprising in view of its origins in the stimulus-response theory of psychology, is that it tends to miss the strategic aspects of human behaviour. It models behaviour as a response to external factors rather than as a choice made among alternatives after weighing the advantages and disadvantages of each.

The philosophical, institutional, and behavioural approaches are all essential to political science, yet each has characteristic limitations. The philosophical approach tends to be abstract and moralistic; institutionalism lacks an account of individual behaviour; and behaviouralism has difficulty modelling individual choices and linking those choices to the institutional context. To some extent, these three approaches can be used in conjunction to help cope with the inherent limitations, and indeed they are rarely practised today in a pure form. Happily, we have moved past the methodological battles of the 1960s, when practitioners of all three approaches sometimes saw the others as enemies rather than potential collaborators. Political philosophers now read the empirical literature, institutionalists draw on behavioural studies, and behaviouralists try to be aware of the institutional context of individual behaviour.

In this tableau, rational choice is not a replacement for existing approaches but a complement with its own characteristic strengths and weaknesses. It must be used in conjunction with the other three; and when properly employed, it helps all the others to work better. What, specifically, does rational choice have to contribute to the mix?

- To complement the philosophical approach, rational choice emphasizes the importance of scarcity and self-interest in human affairs. When political philosophers design institutions to implement justice, equality, tolerance, or some other desirable value, rational-choice analysis will help show why those goals are so hard to attain if they call upon individuals to act in a manner contrary to their own self-interest.
- To complement institutionalism, rational choice can sketch the microfoundations of organization, showing how institutional patterns emerge from individual pursuit of self-interest within defined settings. This merger of perspectives is already far advanced under the heading of neo-institutionalism. Sophisticated institutionalism today draws freely on rational-choice analysis of individual decision-making.
- To complement behaviouralism, rational choice offers an emphasis on conscious choice in pursuit of self-interest. The theory of human nature employed in rational choice makes it possible to construct deductive models of behaviour to be tested against the empirical data that the behaviouralists are so adept in collecting. A natural marriage of convenience is now being consummated by researchers in voting behaviour and other traditional areas of behavioural concern.

This view of rational choice is admittedly a restrained one. In contrast, some proponents have promoted it as a veritable revolution in political science, the first and only way of studying politics scientifically. But we should be cautious about such claims because we have heard them before, namely from the proponents of the behavioural revolution forty years ago.

In fact, the claim to be founding a new science is a recurrent theme in all the social sciences. It dates from the *loi des trois états* (law of the three stages) propounded by Auguste Comte, who invented the term 'sociology.'[20] According to Comte, his new science, which he variously called 'social science,' 'sociology,' and 'positive politics,' was the first genuinely scientific account of human behaviour, and all other knowledge had to be consigned to the garbage can of outmoded 'metaphysical' and 'theological' ideas.[21] Ever since Comte first set the agenda of social science, his megalomania lingers on in sociology, political science, and other disciplines. The result is a recurrent tendency to label methodological innovations not just as potentially useful additions to the study of human behaviour, but as revolutionary breakthroughs that replace rather than complement existing knowledge.

From the standpoint of scholarship, one of the worst features of Comtean megalomania is not just the exaggerated claims it makes, but also the exaggerated reactions it engenders. Proponents of older methodologies naturally rush to denounce arrogant newcomers and point out their defects – no doubt a necessary task. But ultimately the progress of knowledge depends on finding out what new approaches can do rather than on harping on what they cannot do. That is the spirit I have tried to adopt in this book – to work with rational choice, seeing what I can get out of it that makes Canadian politics more understandable. The result is a series of narrowly focused applications or case studies. In the last chapter, I will return to the topic of how they complement, rather than displace, older approaches.

2
Game Theory

Game theory is a branch of mathematics involving models of situations in which outcomes are interdependent. That is, no player can determine the result by himself, for the outcome arises from the interplay of the choices made by all the players. A game model requires the following elements: players, rules of the game, strategies, payoffs, and a solution (or solutions).

- Players are assumed to be rational actors armed with ordinal or cardinal utility functions. There must be two or more players.
- Rules of the game define the limits of action – what can and cannot be done in the game.
- Strategies are the choices that the players can make within the rules of the game. A strategy is a complete set of choices from beginning to end of the game. For example, if a player can make three different decisions, and at each decision he can choose between two alternative, he has $2^3 = 8$ different strategies for the whole game.
- Payoffs are the outcomes that accrue to players depending on the choice of strategies they and their opponents make. Payoffs may be either ordinal or cardinal.
- A solution is the set of payoffs arising from the strategies that rational players would choose under the rules of the game. Sometimes there are multiple solutions. Indeed, sometimes there are multiple *solution concepts;* that is, more than one line of reasoning that rational actors might employ. Finding solutions and examining the merits of different solution concepts is what game theory is all about.

TABLE 2.1
Example of a Game in Normal Form

		Column Player	
		Strategy a	Strategy b
Row Player	Strategy A	1, −1	2, −2
	Strategy B	3, −3	4, −4

Let's illustrate this abstract terminology with concrete examples, starting with a purely numerical example, depicted in so-called *normal* or *strategic* form (see Table 2.1).

The numbers in the cells of the matrix are the payoffs that the players get. They are cardinal numbers, but you don't have to worry about what they measure; they could be in any units whatsoever. By convention, Row Player's payoffs are always written first and Column Player's second. (When first studying game theory, I found it convenient to use the mnemonic RC, as in Roman Catholic, to remember the order.)

This game is stacked against Column Player because his payoffs are always negative. Yet he can do better or worse, with scores ranging from as high as −1 to as low as −4. As a rational actor, he would like to lose as little as possible. Row Player always wins something, but he also can do better or worse for himself, with scores ranging from 1 to 4. The exercise is like duplicate bridge, where your score depends not on the cards you get but on how much you can squeeze out of whatever cards you are dealt.

What strategic choice will each player make? Remember that, under the assumption of perfect information, each player sees the whole gameboard. Each player also knows that the other player sees the whole gameboard, and each knows that the other knows that the opponent sees the whole board, and so on.

Row Player should play *B* because *B* always gives Row Player a higher payoff than *A*. If Column Player chooses *a*, Row Player gets 3 for choosing *B*, in comparison to 1 for playing *A*. If Column Player chooses *b*, Row Player gets 4, in comparison to 2 for playing *A*. So *B* is always better than *A*. In the language of game theory, we say that *B dominates A* because it is an unconditionally superior choice.

Column Player also has a dominant strategy, namely *a*. If Row Player chooses *A*, Column Player gets –1, as compared to –2 for playing *b*. If Row Player chooses *B*, Column Player gets –3, as compared to –4 for playing *b*. So *a* is always better than *b*. Because Row Player always chooses *B* and Column Player always chooses *a*, the solution of the game is the pair of payoffs (3, –3) produced by the strategy pair *Ba*. The solution is printed in boldface in the lower-left cell of Table 2.1.

Both players in a game do not necessarily have dominant strategies, but there will be a solution even if only one player has a dominant strategy. Consider Table 2.2, which is similar to Table 2.1 except that the payoffs in the lower row have been reversed.

For Row Player, *B* is still a dominant strategy, but matters are not so simple for Column Player. If Row Player plays *A*, Column Player would prefer to play *a* rather than *b*, in order to get –1 rather than –2. But if Row Player plays *B*, Column Player would prefer to play *b* rather than *a*, in order to get –3 rather than –4. So Column Player does not have a dominant strategy. Nonetheless, the game has a solution. Knowing that Row Player will play his dominant strategy, *B*, Column Player will choose *b* in order to get –3 rather than –4. The solution, therefore, is (3, –3), which has the same value to both players as in the first model, but now represents the strategy pair *Bb*.

Games can also be depicted in *extensive* form, as shown in Figure 2.1, which contains exactly the same information as the game in Table 2.1.

In this *game tree*, the dots are known as *nodes*. Those from which branches emerge are choice nodes; they represent points where one player or the other selects between alternative strategies. Those dots at the bottom are called terminal nodes; they represent the possible outcomes of the game. The *branches* at each choice node represent the strategic choices available to the player at that point. In the payoffs

TABLE 2.2
Second Example of a Game in Normal Form

		Column Player	
		Strategy a	Strategy b
Row Player	Strategy A	1, –1	2, –2
	Strategy B	4, –4	**3, –3**

FIGURE 2.1
Example of Game in Extensive Form

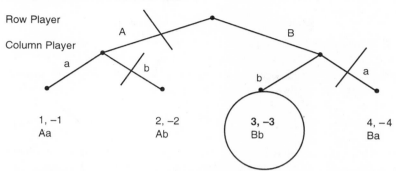

printed below the terminal nodes, the convention is that the payoff printed first is that of the player whose move is depicted first – Row Player in this case.

The solution is found by a reasoning process known as *backwards induction*, which starts at the bottom of the game tree and works upward. Put yourself in the position of Column Player, who has two choice nodes: one at the bottom of Row Player's branch A, and the other at the bottom of branch B. In each instance, Column Player has a choice between a and b. At the first choice node, Column Player will choose a over b, because that gives a payoff of –1 rather than –2. At the second choice node, Column Player will also choose a over b, because that gives a payoff of –3 rather than –4. For this reason, both branches representing strategy b have a bar through them to indicate that Column Player will not choose them.

Now go up the tree to Row Player's choice node, keeping in mind that Row Player knows how Column Player will reason. Because Column Player chooses a in both cases, Row Player has an effective choice between Aa and Ba. The former gives Row Player 1, the latter 3, so he will choose B, resulting in strategy pair Ba and outcome (3, –3) as the solution. Therefore, branch A is marked with a bar because a rational Row Player will never choose it. Visually, the solution is reached by starting at the first choice node and following the only unobstructed (unbarred) path to the end of the game.

Theoretically, any game can be depicted in either normal (tabular) or extensive form. However, in practice, normal form lends itself well to games in which players make simultaneous choices, and extensive form is more convenient for games in which players make sequential choices

– that is, take turns in making their decisions. Both will be used in the rest of this book, so take a minute to be sure you understand what has been done thus far.

Some Refinements

A few additional points need to be made in order to help you understand the material in the remainder of the book.

N-Person Games

For simplicity, the exposition of game theory always starts with two-person games, but more than two players may be involved. Most political situations have to be modelled with n-person games, often with very large numbers of players, as in national elections. Although there are many similarities between n-person and two-person games, they differ in one fundamental respect – the formation of coalitions. In n-person games in which cooperation is allowed between players, coalitions can prosper through the distribution of the payoffs expropriated from players excluded from the winning coalition. In contrast, two-person games force the players to deal only with each other; players may derive mutual benefit from cooperation, but not at the expense of excluded third parties.

Variable-Sum Games

All the examples to this point have been zero-sum games, so-called because the two payoffs in every cell add up to zero. Zero-sum games are games of perfect competition. What one player wins, the other player loses; there is no common ground. In fact, the common name zero-sum is a bit misleading. A better name would be constant-sum, because the key fact is that the two payoffs in all cells add up to the same value, whether or not it is zero. For example, if we add 5 to all the payoffs in Table 2.1, as shown below in Table 2.3, the only thing that changes is that the sum of the payoffs in each cell is now 10 rather than zero. Each player still has the same dominant strategy as in Table 2.1, and the solution is consequently still the same strategy pair. As in all constant-sum games, whatever is gained by one player is entirely at the expense of the other.

TABLE 2.3
Example of a Constant-Sum Game

		Column Player	
		Strategy a	Strategy b
Row Player	Strategy A	6, 4	7, 3
	Strategy B	**8, 2**	9, 1

Multiple Choices

For ease of presentation, all models thus far have featured players with only two strategic alternatives at each choice. Game theory, however, is not limited to such simple situations. Consider the matrix in Table 2.4, where each player has three options from which to choose.

In Table 2.4, neither player has a dominant strategy, but the game can still be solved by a procedure known as the iterated deletion of dominated strategies. A *dominated* strategy is one that will never be chosen. If there are only two strategies and one is dominant, the other must be dominated; but things are more complex when there are more than two strategies. In the matrix in this table, strategy A is dominated by B, but B does not dominate C (Row Player would prefer to play C if Column Player plays c). Similarly, Column Player does not have a dominant strategy but does have a dominated strategy, namely c, whose results are always inferior to those obtained by playing b. Since the dominated strategies A and c will never be chosen, we can delete them, leaving the 2 × 2 matrix of Table 2.5 (the reduced game).

TABLE 2.4
Example of a Game with Multiple Choices

		Column Player		
		a	b	c
Row Player	A	1, −1	2, −2	3, −3
	B	**4, −4**	5, −5	6, −6
	C	1, −1	0, 0	7, −7

TABLE 2.5
Reduced Game

		Column Player	
		a	b
Row Player	A	1, −1	2, −2
	B	**4, −4**	5, −5

In the reduced game, each player has a dominant strategy, namely B for Row Player and a for Column Player. Hence, the solution (4, −4) is produced by the strategy pair Ba.

Ordinal Games

All the examples thus far have been games with cardinal payoffs. Later in the book, however, you will see examples of ordinal games, in which the numbers in the payoffs represent preference rankings rather than measured quantities. Sometimes they can be solved in the same way as cardinal games, by looking for dominant strategies, as in the example of Table 2.6.

To interpret the payoffs, think of them as statements about preferences: $4 > 3 > 2 > 1$. Row Player's first preference among possible outcomes would be strategy pair Bb, yielding the outcome in the lower-right corner of the matrix, (4, 3); and Column Player's first choice would be Ba, yielding (3, 4) in the lower-left corner. Each player in the game has a dominant strategy. Row Player always plays B because $3 > 1$ and $4 > 2$, while Column Player always plays a, because $2 > 1$ and $4 > 3$. The solution, therefore, is Ba, in which Row Player gets his second preference and Column Player gets his first.

TABLE 2.6
Example of an Ordinal Game in Normal Form

		Column Player	
		Strategy a	Strategy b
Row Player	Strategy A	1, 2	2, 1
	Strategy B	**3, 4**	4, 3

Mixed-Strategy Solutions

All the examples we have looked at thus far have solutions in pure strategies. Each player has an unambiguously rational choice of one and only one strategy, so the solution is the outcome of a strategic pair. However, there are many games where the rational choice for each player is to play a random mixture of two or more options in specified proportions. Consider the old schoolyard game of matching pennies. You and I each simultaneously toss a penny. If they match when they come down (both heads or both tails), I keep my penny and win yours; if they do not match (one is heads and the other is tails), you keep your penny and win mine. (Tossing the coins makes it a game of chance, but it could just as well be a game of deliberate strategy if the players secretly chose heads or tails and then simultaneously revealed their choice, as in that other schoolyard favourite, rock-scissors-paper.) Matching pennies is modelled in Table 2.7.

Neither player has a dominant strategy in this zero-sum game, so there is no solution in pure strategies. It is easy to see that any player who chooses only one strategy is in trouble. If I always play heads, you can always win by playing tails; and if I always play tails, you can always win by playing heads. Similarly, if you always play heads (or tails), I can always win by matching you.

Intuitively, it is obvious that each player should randomly choose heads half the time and tails half the time. If we each play that way, we will each win half the time and lose half the time, for a payoff of (0, 0). In technical language, we would say that the solution in mixed strategies is for each player to play heads with a probability of $p = 0.5$ and tails with a probability of $1 - p = 0.5$, for a value of 0 to each player. Since tossing the coins gives precisely that result, it is understandable how the game came to incorporate the toss as part of the rules.

In this case, the solution is obvious, but only because the payoffs are perfectly symmetrical. How can we demonstrate that one should play

TABLE 2.7
Matching Pennies

		You	
		Heads	Tails
Me	Heads	1, −1	−1, 1
	Tails	−1, 1	1, −1

TABLE 2.8
Matching Pennies with Algebraic Probabilities

| | | | You | |
			(q) Heads	(1–q) Tails
Me	(p)	Heads	1, –1	–1, 1
	(1 – p)	Tails	–1, 1	1, –1

heads and tails in a 50:50 ratio? The general solution, of which this is one example, was first demonstrated in 1926 by John von Neumann in the Minimax Theorem.[1] We can explain it in the following case.

I do not have a single winning strategy, but I want to do as well as I can for myself. Since this is a zero-sum game, your win comes at my expense. Hence, doing as well as I can for myself means forcing you to do as badly as is within my power. To put it another way, I want to minimize the maximum that you can obtain (hence, the term minimax).

I minimize your maximum if I can find a mixture of strategies such that you cannot get any better results, no matter what you choose. That task implies that I mix my strategies so as to equalize the results for you, so that you get the same average payoff by playing heads as by playing tails. If that is the case, you have no way to improve, which is what I am aiming at.

Table 2.8 shows the game matrix with algebraic expressions added for the probabilities that we wish to compute.

Now we bring in the logic of expected value. If I equalize your results, that means the expected value you receive from playing heads will equal the expected value you receive from playing tails, as expressed in Equation 1:

(1) $EV(H) = EV(T)$

in which EV is expected value, H is heads, and T is tails.

The expected value you get from playing heads equals the sum of your payoffs for playing heads against my heads and my tails; similarly, the expected value you get from playing tails equals the sum of the payoffs you get from playing tails against my heads and tails. Thus we get Equations 2, 3, and 4:

(2) $EV(H) = EV(HH) + EV(HT)$

(3) $EV(T) = EV(TH) + EV(TT)$

(4) $EV(HH) + EV(HT) = EV(TH) + EV(TT)$

The term $EV(HH)$ is equal to the payoff you get for HH, which is -1, multiplied by the probability that I play H, which is p. Similarly, the term $EV(HT)$ is equal to the payoff you get for HT, which is 1, multiplied by the probability that I play T, which is $1-p$. After $EV(TH)$ and $EV(TT)$ are expressed in similar fashion and all terms are substituted into Equation 4, we get Equation 5, which is then simplified and solved for p and $1 - p$ in Equations 6 through 10:

(5) $(-1)p + 1(1 - p) = 1p + (-1)(1- p)$

(6) $-p + 1 - p = p - 1 + p$

(7) $1 - 2p = 2p - 1$

(8) $-4p = -2$

(9) $p = 0.5$

(10) $1 - p = 0.5$

What this means is that, if I wish to achieve my minimax point by equalizing your outcomes, I should play heads and tails each with a probability of 0.5. A separate set of calculations – Equations 11 through 20 – is necessary to find q and $1 - q$. In this perfectly symmetrical game, p and q happen to be the same, but that is not generally true. The calculation of q proceeds by setting up equations with my payoffs, just as p was calculated by setting up equations with your payoffs.

(11) $EV(H) = EV(T)$

(12) $EV(H) = EV(HH) + EV(TH)$

(13) $EV(T) = EV(HT) + EV(TT)$

(14) $EV(HH) + EV(TH) = EV(HT) + EV(TT)$

(15) $1q + (-1)(1 - q) = -1q + 1(1 - p)$

(16) $q - 1 + q = -q + 1 - q$

(17) $2q - 1 = 1 - 2q$

30   Game Theory and Canadian Politics

(18) $4q = 2$

(19) $q = 0.5$

(20) $1 - q = 0.5$

Like me, you too should play heads half the time and tails half the time.

A Real-World Example

Although this is a book on Canadian politics, I hope the reader will tolerate, perhaps even relish, an example drawn from baseball. I want to illustrate the concept of a solution in mixed strategies with a concrete example using real-world data, but I have not been able to find an appropriate one from Canadian politics.

Zero-sum games, although they are the conceptual foundation of game theory, are rather limited in their application to politics, because there is an element of cooperation in most political situations. Political rivals usually have ways to make each other simultaneously better or worse off, rather than simply win at the expense of the other. 'Win-win' and 'lose-lose' are more common than the 'win-lose' situations of zero-sum models.

One zero-sum, or more properly constant-sum, situation is the electoral struggle for seats in the House of Commons or provincial legislatures. Given a constant number of seats, a riding won by one party cannot be won by another party. But most Canadian elections feature contests between more than two parties and typically include a choice of many electoral strategies; so any mixed-strategy model would have to feature several players with several choices to make, and the mathematics would become far more complex than appropriate for this book.

Although baseball is a complex team sport, the focus of attention is the duel between pitcher and batter. This central contest can readily be modelled as a two-person, zero-sum game with a binary choice of strategies for each player. The pitcher can throw with either the right or the left hand, and the batter can hit from either the right or the left side. There are thus four possible pitching-batting combinations, as shown in Table 2.9.

Table 2.9 does not apply to a single encounter. Pitchers have only one good throwing arm, and batters are not allowed to switch sides during their turn at the plate. However, the model does apply in a sense to

TABLE 2.9
Pitching–Batting Combinations

		Pitcher	
		Throws left	Throws right
Batter	Hits left	L vs. L	L vs. R
	Hits right	R vs. L	R vs. R

managers assembling a lineup. All batting orders contain a mixture of right- and left-handed hitters, and all pitching rosters contain a mixture of right- and left-handed pitchers. During the season a pitcher can expect to face batters standing on both sides of the plate, and a batter can expect to face pitchers throwing from either side of the mound.

However, we would not expect the model to be a good representation of any particular team. Professional baseball rosters are relatively small, containing about 10 pitchers and 15 other players for a total of 25. The pitchers on a team, since all they do is pitch, might be balanced between left and right according to some mathematical principle; but hitters (except for designated hitters in the American League) have to play a defensive position and be able to run the bases. Moreover, because players under long-term contracts are not freely available to be hired and fired, managers work with a limited pool of candidates in any particular year. Hence, any roster will have to balance a number of athletic and economic considerations, and predictions cannot be made solely from a model of batting success.

The model, however, can be tested at a higher level of aggregation, in which there is a sort of statistical contest being waged by hitters as a group against pitchers as a group. In the 1995 season, for example, 550 pitchers faced 583 batters in 133,621 at-bats.[2] We could reasonably expect a game-theory model to be predictive at this level. That is, it could be used to predict the proportions of left- and right-handed batters and pitchers in the whole population of major-league baseball. Biologists have done similar exercises with mixed-strategy models to predict population polymorphism; for example, the proportions of males and females in a species.[3]

Even the most casual baseball fan knows that right-handed batters tend to do better against left-handed pitchers, while left-handed batters

32 Game Theory and Canadian Politics

do better against right-handed pitchers. We have all seen games held up while one manager sends in a pinch hitter to get the batter's opposite-side advantage, and the other manager responds by changing pitchers to restore the hurler's same-side advantage. The reason for the batter's opposite-side advantage has to do with the behaviour of breaking-ball pitches. A natural curve ball, that is, one thrown with outward rotation of the pitcher's arm, breaks to the opposite side – to the left for a right-handed pitcher and to the right for a left-handed pitcher. A curve ball also drops and acts as a change of pace because it is slower than a fastball. All these factors fool the batter to some extent, but a curve ball moving toward you is easier to hit than one moving away from you. Robert K. Adair explains this phenomenon in *The Physics of Baseball*: 'Batting against the curve ball, the batter tends to swing too quickly at the relatively slow pitch, and he tends to underestimate the in-out curve deviation. These errors tend to add up for out-curves but cancel for in-curves. Hence, the in-curves may be a little easier to hit, accounting for the small advantage batters have when they face a pitcher throwing from the opposite side.'[4] Data compiled and published by STATS, Inc., can be used to develop and test a model of the contest between pitchers and hitters. I use batting average as a payoff indicator. Other measures exist, such as slugging average, on-base percentage, and even more refined measures of run production. However, batting average is the most widely reported, best known measure.[5]

Table 2.10 summarizes the batting results for the 1995 season after removing the at-bats of pitchers in the National League (because pitch-

TABLE 2.10
Batting Results, 1995*

	Batter	Pitcher	At–bats	Hits	Batting average
1	L	L	8,360	2,144	.2565
2	L	R	30,479	8,559	.2808
3	L	L&R	38,839	10,703	.2756
4	R	L	20,609	5,713	.2772
5	R	R	48,361	12,746	.2636
6	R	L&R	68,970	18,459	.2676
7	S	L	7,165	1,927	.2689
8	S	R	18,647	5,030	.2697
9	S	L&R	25,812	6,957	.2696
10			133,621	36,119	.2703

* After removing at-bats of National League pitchers

TABLE 2.11
Pitcher–Batter Game, 1995 (Switch-hitters Merged)

| | | Pitch | |
		Left	Right
Bat	Left	.2565, −.2565	.2766, −.2766
	Right	.2751, −.2751	.2636, −.2636

ers do not bat at all in the American League). I used the simplifying procedure of counting switch-hitters as left-handed batters when they faced right-handed pitchers and as right-handed batters when they faced left-handed pitchers. This means combining lines (8) and (2) and lines (7) and (4) from Table 2.10 in order to produce Table 2.11. The purpose is to enable construction of a 2 × 2 model, which is the easiest to solve and interpret.

Inspection of Table 2.11 shows there is no solution in pure strategies for the 1995 Pitcher-Batter Game. Neither side has a dominant strategy; it all depends on what the other side is doing. If the pitcher is left-handed, you will do better on average by sending out a right-handed batter, and vice versa. Hence, we must look for a solution in mixed strategies, which means that each player should play a random combination of strategies in prescribed proportions. The problem is to compute the proportions – that is, to find the prescribed probability p of batting left and $1 - p$ of batting right, and the prescribed probability q of pitching left and $1 - q$ of pitching right.

The solution is similar in principle to that for the game of Matching Pennies described above. Each player wants to select probabilities such that his opponent's expected payoff is the same no matter which strategy the opponent selects. In other words, batters want to mix their likelihood of batting right and left in such proportions that it does not matter whether the pitcher throws right or left. Similarly, pitchers want to mix their probabilities of throwing right and left in such proportions that it does not matter whether hitters bat right or left.

Algebraically, this solution implies the following set of equations for calculating p and $1 - p$ in the 1995 Pitcher-Batter Game:

Let EV = Expected Value

$EV(\text{Pitch } L) = EV(L \text{ vs. } L) + EV(L \text{ vs. } R)$

$EV(\text{Pitch } L) = -0.2565p + (-0.2751)(1 - p)$
$EV(\text{Pitch } R) = EV(R \text{ vs. } L) + EV(R \text{ vs. } R)$
$EV(\text{Pitch } R) = -0.2766p + (-0.2636)(1 - p)$
$EV(\text{Pitch } R) = EV(\text{Pitch } L)$
$-0.2565p + (-0.2751)(1 - p) = -0.2766p + (-0.2636)(1 - p)$
$p = 0.364$
$1 - p = 0.636$

If there had been a manager in charge of all hitters for 1995, he should have chosen lineups such that players batted left 36.4 per cent and right 63.6 per cent of the time. With this combination, it would not have mattered whether the roster of batters faced a right- or left-handed pitcher; the expected payoff would have been the same. This is the best that such a hypothetical manager could have done. Had he fielded more left-handed batters than the optimum, the defence could have done better by pitching left-handed as a pure strategy. Similarly, had he played more right-handed batters, his hitters could have been victimized by right-handed pitchers.

The calculation of q and $1 - q$ proceeds along similar lines:

$EV(\text{Bat } L) = EV(L \text{ vs. } L) + EV(RL \text{ vs. } R)$
$EV(\text{Bat } L) = 0.2565q + 0.2766(1 - q)$
$EV(\text{Bat } R) = EV(R \text{ vs. } L) + EV(R \text{ vs. } R)$
$EV(\text{Bat } R) = 0.2751q + 0.2636(1 - q)$
$EV(\text{Bat } L) = EV(\text{Bat } R)$
$0.2565q + 0.2766(1 - q) = 0.2751 + 0.2636(1 - q)$
$q = 0.411$
$1 - q = 0.589$

A hypothetical manager in charge of all pitchers in 1995 should have had 41.1 per cent of at-bats pitched by left-handers and 58.9 per cent by right-handers. The offence could have penalized any departure from this policy by converting to the pure strategy of always batting right or left, in coordination with the shortage of pitchers (that is, if there are not enough left-handed pitchers, always bat left, and so on).

TABLE 2.12
Comparison of Actual and Predicted Values from the Pitcher–Batter Game, 1995 (Switch-hitters Merged)

	Bat	Pitch	Per Cent	
			Actual	Predicted
1	L	–	43.0	36.4
2	R	–	57.0	63.6
3	–	L	27.0	41.1
4	–	R	73.0	58.9

The empirical test of the model is to compare its predictions to the actual proportions calculated from Table 2.11. The comparison is presented in Table 2.12.

The model underpredicts the percentage of left-handed batters somewhat (36.4 per cent vs. an actual figure of 43.0 per cent) and overpredicts the percentage of left-handed pitchers even more (41.1 per cent vs. an actual figure of 27.0 per cent). Explaining these discrepancies leads us into the realm of human biology. About 14 per cent of males are born left-handed.[6] Left-handed people experience earlier mortality, but that is not a significant factor at the age at which professional baseball is played.[7] Hence, the pool of potential professional players is about 14 per cent left-handed. This finding suggests that baseball teams would hire more left-handed pitchers if they could find them; but there are just not enough available. Indeed, considerable anecdotal evidence reveals that left-handed pitchers are always in demand and can command premium salaries.[8]

If biological realities explain the shortage of left-handed pitchers, how do we explain the actual surplus of left-handed batters? Here the logic of game theory comes into play. There is no penalty for having too many left-handed batters because the defensive side does not have enough left-handed pitchers at its disposal. Indeed, given the shortage of left-handed and the surplus of right-handed pitchers, the batting side would actually do better to abandon its mixed strategy and to play nothing but left-handed batters.

In a sense, then, the question is, given the shortage of left-handed pitchers and the surplus of right-handed pitchers, why are there any right-handed batters at all? Why don't managers hire nothing but left-handed batters? One answer may be that the defensive positions of second base, shortstop, and third base favour right-handed throwing because of the body position required to throw to first base. The reasons are not so obvious for the catcher's position, but there too right-handers are almost universal. Because right-handed throwers are needed for many defensive positions, right-handed batting often comes along as part of the package.

If you're a baseball fan, additional questions are probably bubbling up in your mind; and if you're not a baseball fan, you're undoubtedly saying, 'Enough already!' Fans can pursue the matter at greater length in specialized journal articles.[9] Here let me explain why I used this example rather than one drawn from Canadian politics.

First, the algebraic solution in mixed strategies requires cardinal data in order to set up and solve the equations. Some political games have cardinal payoffs (number of seats or percentage of popular vote won by a political party in an election, for example; or the amount of funding secured from the government by an interest group). In many situations, however, the payoffs are ordinal; namely, securing a more preferred over a less preferred alternative, as when a group succeeds in getting a bill amended during its passage through Parliament.

Second, the pitcher-batter duel is a two-person game in which each player has two options, throwing or batting left or right. Although there are some two-person political situations (particularly bargaining situations like the Lubicon conflict described in the next chapter), most involve more than two actors. Also, although there are some political situations in which actors have only two choices (voting yes or no in a referendum or in a legislature, for example), it is more common for players to have several options at their disposal.

I could not find a political situation in which there were only two players with exactly two strategies to pursue, and in which the payoffs could fairly be represented by cardinal numbers. But in the absence of a political example, the baseball example shows how a game-theoretical model can be constructed and tested against real-world data.

3

Stalemate at Lubicon Lake

As we saw in the preceding chapter, the minimax solution of two-person zero-sum games is mathematically elegant and satisfying, but few political situations can be realistically modelled by such games. Hence, two-person zero-sum games are of limited use in political science. However, variable-sum games – those in which the payoffs in the various cells add up to different totals – are applicable to a much wider variety of situations.

Table 3.1 is a simple example of the famous Prisoner's Dilemma game, which is widely used as a model in political science. It is presented here, however, only as an illustration of how variable-sum games are solved. Note that the sum of the two players' payoffs ranges from 6 in the upper-left cell through 5 in each of the two off-diagonal cells to 2 in the lower-right cell; hence the name variable-sum game.

Zero-sum games model situations of pure competition, in which one player's gain is exactly equal to the other player's loss. Variable-sum games, in contrast, model situations in which a player can make an-

TABLE 3.1
Prisoner's Dilemma

		Player B	
		Cooperate	Defect
Player A	Cooperate	3, 3	0, 5
	Defect	5, 0	**1, 1**

other better off even as he improves his own payoff. Usually there is a mixture of cooperation and competition in a variable-sum game. In Table 3.1, both players achieve a payoff of 3 if they cooperate, compared to 1 if they both defect. That is to say, cooperation makes them better off both individually and collectively. But the competitive element remains because one player can get 5 at the expense of the other, who then gets 0.

Looking for dominant strategies, as in a zero-sum game, quickly shows that the equilibrium is in the lower-right cell – mutual defection, with a payoff of 1 to each player – because defection is a dominant strategy for each player. Player A will reason that Player B can either cooperate or defect. If B cooperates, A gets 3 by also cooperating but 5 by defecting. If B defects, A gets 0 by cooperating but 1 by also defecting. In either case, the strategy of defection is dominant over the strategy of cooperation. But since B's situation and reasoning process are exactly analogous to those of A, both players will choose defection, and each will get 1.

In more general terms, the solution concept applying to variable-sum games is known as the *Nash equilibrium*, defined as an outcome over which neither player can improve by making a unilateral move. In Table 3.1, Player A's payoff declines from 1 to 0 if he moves to cooperation while B remains at defection, and likewise for a unilateral move by B. This test will identify Nash equilibrium points even in the absence of dominant strategies.

Any reader who has ever studied economics will be familiar with the concept of the *Pareto optimum* – the outcome which makes everyone as well off as possible without making anyone worse off. The Nash equilibrium and the Pareto optimum may coincide, but they may also diverge, as they do in Table 3.1. Mutual cooperation is the Pareto optimum because (3, 3) is the highest payoff jointly attainable. Each player can get 5 only if the other player becomes worse off by getting 0. But mutual cooperation is not a Nash equilibrium – not stable – because each player would benefit from a unilateral move. If B cooperates, A can improve from 3 to 5 by defecting, and similarly for B if A cooperates.

The relationship is easy to see in the graphic form of the game shown in Figure 3.1. The point (3, 3) is the Pareto optimum because, among the possible outcomes of the game, it lies farthest up and to the right, affording the highest payoffs jointly to both players. The Nash equilibrium (1, 1) is Pareto-inferior in comparison to (3, 3).

FIGURE 3.1
Graphic Representation of Prisoner's Dilemma

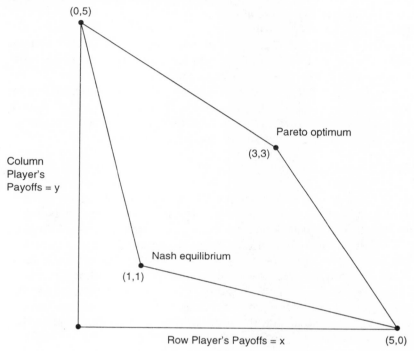

The enduring interest of Prisoner's Dilemma lies precisely in this divergence between the Pareto optimum and the Nash equilibrium. The Pareto optimum is desirable because it is better for both players, but it is not stable because both players have an incentive to move away from it. The Nash equilibrium is stable because neither player has an incentive to move from it, but it is not desirable, at least not in comparison to the Pareto optimum.

Although Prisoner's Dilemma is the most famous variable-sum game, it is not the only one. There is an infinity of human situations involving cooperation and competition, and an infinity of variable-sum games can be devised to model them. The rest of this chapter uses a version of the game known as Deadlock to model a notorious stalemate in Canadian politics – the standoff between the federal government and the

Lubicon Lake Cree band of northern Alberta. The case study shows how use of a formal model can help clarify the motives of parties in conflict and also test their behaviour for logical consistency.

The example uses only ordinal payoffs. In many political situations, it is difficult or impossible to assign cardinal values to outcomes, and yet players have a clearly discernible preference ordering. They know what they want and are capable of ranking the outcomes. Fortunately, the solution concept of the Nash equilibrium, unlike the von Neumann minimax solution used in the preceding chapter, works just as well with ordinal as with cardinal payoffs, because it only requires players to compare possible payoffs and decide which is preferable.

The Lubicon Lake Conflict[1]

Lubicon Lake lies in a vast wilderness of woods and water north of Lesser Slave Lake between the Peace and Athabasca rivers, in the midst of the Treaty Eight area. When the treaty commissions passed through northern Alberta in 1899 and 1900, there were no trails to accommodate such a large party of white men and their supplies. They had to travel by water, making a circle around the Lubicon homeland and stopping at trading posts and missions that the Indians were accustomed to visit. Although the government realized that as many as 500 Indians had not taken treaty, it decided that the Indian title could be considered extinguished and that it was not necessary to send out further treaty parties; the Treaty Eight inspectors were authorized to add individuals to treaty lists when they made their annual rounds.

In 1933, 14 men petitioned the government to create a new reserve at Lubicon Lake, pleading that they did not wish to live at the Whitefish Lake reserve because it was too far away. In 1940, the Indian Affairs Branch gave the Lubicons permission to elect their own chief, but it was too late for them to obtain a reserve directly from the federal government. The Natural Resources Transfer Agreement embedded in the *Constitution Act* of 1930 had transferred public lands from the jurisdiction of the Crown federal to the Crown provincial, so the federal government would have to request land from Alberta after validating the claim.

An Indian agent visited Lubicon Lake in 1940 with a federal surveyor and selected an approximate location for the reserve at the west end of the lake. The surveyor general could not afford to do the survey in 1941 because the Second World War had reduced his budget; but in 1942 the

Indian Affairs Branch requested Alberta to designate 25 square miles west of Lubicon Lake as a probable reserve. After the war ended, the surveyor general turned to the task of surveying the proposed reserves in northern Alberta; but for reasons unknown, the survey at Lubicon Lake was apparently forgotten.

In 1950, the Lubicons again asked for their reserve, touching off several confusing years. Local officials were instructed to ask them if they really wanted a reserve at Lubicon Lake or if they might prefer some other location. These consultations proved unsatisfactory because the Lubicons were still actively trapping and hunting. Only a handful might show up for any particular meeting, and opinion about the location of the reserve varied, depending on who was present.

Alberta, which had been carrying a provisional reserve on its books, issued an ultimatum on 22 October 1953: if Ottawa did not commit itself within 30 days, the province would take the reserve off its books and open the area to oil exploration. Indian Affairs, advised by local officials that Lubicon Lake was an unsuitable location and that some other solution could eventually be found, made no effort to block Alberta's action, and the provisional reserve disappeared.

The Lubicons continued to live for another two decades as they always had, by hunting, fishing, and trapping, until the province began to build an all-weather road from Grande Prairie to Little Buffalo in 1971. Rising world oil prices pointed to the development of Alberta's oil sands, and Premier Peter Lougheed officially announced, 18 September 1973, that the Syncrude project would go ahead.

In 1975, the Indian Association of Alberta tried to register a caveat to about 25,000 square miles lying north of Lesser Slave Lake between the Peace and Athabasca rivers. This effort was on behalf of the 'isolated communities' – Lubicon Lake and half a dozen other groups of Indians in the area – who claimed to be entitled to reserves under Treaty Eight. By attempting to register a caveat, the isolated communities were asserting an unextinguished aboriginal title to a large part of northern Alberta. As a matter of political strategy, they were trying to block Syncrude and other northern oil projects in order to wring concessions from the provincial government. Fearing that it might lose the impending court battle, the government of Alberta had the legislature revise the *Land Titles Act*. The province declared that there were no unextinguished aboriginal rights in Alberta, only unfulfilled treaty entitlements, and the attorney general dared Indian groups to go to court if they thought they could prove the existence of unextinguished aboriginal title.

The coalition of isolated communities fell apart, and the Lubicon band emerged as a political actor in its own right. After Bernard Ominayak was elected chief in 1978, he entrusted the Lubicons' legal strategy to a Montreal lawyer, James O'Reilly, who had been instrumental in achieving the James Bay Agreement in northern Quebec. O'Reilly's James Bay strategy – claiming unextinguished aboriginal title, seeking an injunction to hold up natural resource development, entering into negotiations to achieve a settlement – seemed like a model for the Lubicons. But there was a key difference between the two situations. Before 1975, there had been no land surrenders in northern Quebec, so the native claim to unextinguished aboriginal title had *prima facie* validity. By contrast, Treaty Eight purported to extinguish the Indian title to all of northern Alberta, so the Lubicons had to make out the far more doubtful proposition that the absence of an adhesion by what may or may not have been a Lubicon band in 1899–1900 created an unceded region in the middle of the treaty area, like a hole in the middle of a doughnut.

The Lubicons' avowal of unextinguished aboriginal title posed a fundamental challenge to the Numbered Treaties as a whole. According to the terms of these treaties, Indian bands did not cede specific parcels of land on which they were accustomed to live; they ceded 'all their rights, titles and privileges whatsoever' to the large tract of land described in the treaty. When bands adhered to the treaty after the original signing, they simply agreed to accept the terms of the treaty, including the surrender of land rights in the described area; they did not bring new territory in with them.

The minister of Indian affairs accepted the Lubicon claim for negotiation as a treaty entitlement, and in January 1982 a federal-provincial meeting considered the issue. But to negotiate for a reserve within Treaty Eight would undercut the Lubicons' position that they still possessed aboriginal title because they had never adhered to Treaty Eight. O'Reilly, therefore, turned to the Alberta Court of Queen's Bench to request an injunction against resource development. He asked for complete cessation of activity in a 'reserve area' of 900 square miles around Lubicon Lake, and a reduced level of activity in a surrounding area of 8500 square miles called the 'hunting and trapping territory.' The theory behind the claim was that unrestricted resource development posed imminent danger to the Lubicons' land rights.

Justice Gregory Forsyth declined to grant the injunction, holding that 'damages would be an adequate remedy to the applicants in the event they were ultimately successful in establishing any of their positions advanced.'² The Lubicons would first have to prove that their aboriginal title still existed, then seek damages if they could show they had suffered loss. The Alberta Court of Appeal upheld the lower court's decision, and in March 1985 the Supreme Court of Canada refused leave to appeal.

This defeat marked the end of the Lubicons' attempt to use the courts to achieve their objectives. Their supporters say 'it had become obvious they could in all likelihood never get justice through this route,'³ whereas sceptics think they realized the weakness of their aboriginal rights claim and had never seriously intended to litigate it. During *Ominayak v. Norcen* they had already turned to the churches, resulting in a letter from the World Council of Churches to Prime Minister Trudeau accusing the Alberta government of genocide. The Alberta government referred the letter to the provincial ombudsman, who reported in August 1984 that he had found 'no factual basis' for the charges. Bernard Ominayak then released a statement attacking the ombudsman's report. It was indoor-outdoor political theatre, with NDP members of the Alberta legislature and the House of Commons recycling the news stories created by ecclesiastical denunciations of the government.

The political situation changed markedly with the election of a Progressive Conservative majority in Parliament in September 1984. The new minister of Indian affairs, David Crombie, met personally with Ominayak and then appointed E. Davie Fulton, one-time federal minister of justice and member of the Supreme Court of British Columbia, to investigate the case. Fulton spent a great deal of time with the Lubicons and ended up virtually as their advocate. Perhaps leery of the prospect of a report relatively favourable to the Lubicons, the province offered in December 1985 to transfer 25 square miles to the band if it would drop all litigation. But the Lubicons dismissed this offer out of hand because a reserve of 25 square miles corresponded to their band size in 1940 – 127 members – whereas they had grown much larger in the meantime and wanted a correspondingly larger reserve. A quantitative issue such as the size of the band may seem like an ideal subject for negotiation and compromise, but for the Lubicons it was a matter of principle. Their theory of aboriginal rights asserted that they were not yet bound by Treaty Eight because they had never adhered to it. To accept any external definition of their size would undercut their claim to continued possession of aboriginal title.

The prime minister then appointed a new negotiator, Roger Tassé, a retired deputy minister of justice and a main architect of the Canadian

Charter of Rights and Freedoms. Bilateral discussions between Tassé and the Lubicons began on 16 June 1986, but the Lubicons broke off the talks when the federal government disputed the band size claimed by the Lubicons and denied their theory of continuing aboriginal title.

With the failure of negotiations, the Lubicons began to put more emphasis on their attempts to influence public opinion. They had filed a complaint before the United Nations Human Rights Committee in 1984 and had also threatened to disrupt the 1988 Calgary Winter Olympics by asking museums around the world not to participate in the Glenbow Museum's Indian exhibition, 'The Spirit Sings.' In August 1986 Ominayak, his political adviser Fred Lennarson, and some Lubicon elders went on a tour of seven European countries to generate support. The boycott had some success, since a number of museums declined to loan artifacts to 'The Spirit Sings,' but it did not prevent the Glenbow from mounting the exhibit. Lubicon supporters also picketed the cross-country Olympic torch relay sponsored by Petro-Canada.

In October 1987, the federal government appointed a new negotiator, Calgary lawyer Brian Malone. A meeting involving two cabinet ministers was set up in an attempt to get movement before the Winter Olympics opened in February 1988. But the meeting got nowhere, and the minister of Indian affairs, Bill McKnight, gave the Lubicons an ultimatum: return to the table within eight days, or the federal government would take further steps toward a legal resolution.

In February 1988, McKnight asked Alberta to provide land for a reserve and made scarcely veiled threats to sue the province if it did not quickly comply. The McKnight formula would have produced a reserve larger than Alberta had previously been willing to grant but smaller than the Lubicons were demanding. After some fruitless efforts to mediate by Alberta premier Don Getty, McKnight made good on his threat to litigate. On 17 May 1988, federal lawyers filed a statement of claim in the Alberta Court of Queen's Bench demanding that the province yield land for a reserve according to the McKnight formula. Alberta and the Lubicons were joined as defendants in a strange reversal of the litigation of the early 1980s.

It quickly became evident that the Lubicons would do everything possible to avoid testing their claim in court. In early June, Bernard Ominayak began to say openly that the Lubicons were ready to 'assert jurisdiction' – to assume governmental control over their traditional territory in validation of their claim that they had never relinquished their aboriginal title. Ominayak announced that, if an agreement was not reached by 15 October 1988, there would be a blockade of roads onto oil-producing lands. O'Reilly read a statement in court that the Lubicons were asserting jurisdiction and would not participate in any further judicial proceedings.

Events now moved quickly in elaborate choreography. The Lubicons set up their blockade on 15 October; the province secured an injunction against it; and early on the morning of 20 October, heavily armed RCMP officers took down the blockade and arrested 27 people. Getty and Ominayak, who had been in contact behind the scenes, met two days later in the little town of Grimshaw. The same day, Getty agreed to sell the federal government 79 square miles with mineral rights, and another 16 square miles without mineral rights, for a reserve. The total of 95 square miles conformed to the Lubicons' own count of their membership.

But no settlement was possible without the agreement of the federal government. Negotiations began well when Ottawa accepted the 95-square-mile reserve but broke down on 24 January 1989. The biggest single issue was compensation. Maintaining they had a 'comprehensive' claim based on aboriginal rights, the Lubicons demanded compensation from the federal government for failure to extinguish their aboriginal title in 1899. The amount owed – $167 million according to one calculation – would compensate the Lubicons for various federal benefits that they had allegedly not received since 1899 because they had no reserve. The Lubicons were willing to negotiate the amount of compensation but not the principle that something had to be paid. The federal government, on the other hand, viewed this as only a 'specific' claim based on treaty entitlement. It recognized the Lubicons' right to a reserve, and it was willing to pay to set up the reserve – $45 million, according to its calculations. But it refused to pay a general amount for extinguishment of aboriginal title, because in its view aboriginal title had been extinguished all over northern Alberta with the signing of Treaty Eight. If it deviated from this principle in the Lubicon case, it might be forced to regard many other current and potential claims across Canada as 'comprehensive' (aboriginal title) rather than 'specific' (treaty entitlement) claims.

After the breakdown of negotiations, the Lubicons continued to assert jurisdiction by delivering an ultimatum to oil companies: obtain permits from us or shut down your operations. On 1 December 1989, Petro-Canada and Norcen shut in 20 wells rather than make an issue of it. The Lubicons also tried to keep the huge Daishowa pulp mill from cutting any trees in their 'hunting and trapping territory,' which forms part of the forest management area granted to Daishowa by the province. Their attempt to organize a boycott of Daishowa led to a new round of litigation, which, as of 1997, was headed for the Supreme Court of Canada.[4]

Meanwhile, the solidarity of the Lubicons began to break up. Only days after the failure of negotiations, Brian Malone was contacted by Lubicon band members who favoured forming another band and negotiating a separate settlement. On 28 August 1989, the Department of Indian Affairs recognized a new Woodland band of about 350 members, including 117 names previously on the Lubicon list. About 30 of these had been expelled by the Lubicons; the rest seem to have left voluntarily. The new band was an amalgamation of Lubicon dissenters with Indians from nearby 'isolated communities,' such as Cadotte Lake, that had not previously been recognized as bands.

The Woodland leaders immediately began to negotiate a specific claim for a reserve at Cadotte Lake, resulting in an agreement in principle on 26 March 1990. Its terms were similar to the final offer that the Lubicons had rejected. Meanwhile, the federal position is that it will extend similar treatment to any other bands in the isolated communities area that want to enter negotiations. In November 1991, it recognized another band, the Loon River Cree, bordering on the Lubicon claim.[5] If the federal policy is successful, it may lead to further defections from the Lubicons' ranks; and as their numbers continue to drop, their claim to a large reserve and full-scale settlement will seem less plausible.

The Lubicon dispute has been so bitter and protracted because it pits two different views of Indian land rights against each other. Canada refuses to recognize any aboriginal claim because it insists that Treaty Eight has extinguished aboriginal title. The Lubicons say that, because their ancestors did not sign the treaty, they still possess unextinguished aboriginal title. Canada will find it difficult ever to concede this point because there are other groups of Indians in the prairie provinces, as well as in other regions of the country, who claim that they also have never adhered to the local treaty. There is no problem in giving the Lubicons, and other groups in their situation, a treaty entitlement; but to recognize the continuing existence of unextinguished aboriginal title would threaten to upset the entire treaty system.

Modelling the Conflict[6]

Let us attempt to represent the last stage of the Lubicon dispute using a variable-sum game model. We will not attempt to measure the payoffs with cardinal numbers; it will be sufficient to use ordinal payoffs, which rank preferences without measuring them precisely.[7] Both the Lubicon band and the federal government will be considered as rationally pursuing their own interests and ranking their preferences in a consistent way. Conflict arises because the two players have different interests and different priorities.

The Lubicons want a settlement based on their theory of aboriginal rights, which would give them, in addition to a reserve of 95 square miles, a large amount of money, including perhaps $100 million compensation for extinguishment of aboriginal title, and many other rights, including self-government and management of wildlife in the region. The main alternative at this stage is a reserve of 95 square miles based on the theory of treaty entitlement, with less cash and fewer other benefits. At earlier stages in the dispute, there were even less generous options with smaller reserves; but these dropped out of consideration with the events of fall 1988. The Lubicons will not accept the treaty option, even though it is now more generous than ever before, and they are willing to remain without a treaty settlement in hopes of obtaining the more lucrative aboriginal rights settlement. Note, however, that their acceptance of the status quo hinges upon the belief that it is only a temporary stage preparatory to obtaining the bigger prize. It would be irrational to prefer the status quo to the treaty settlement if the status quo were to be permanent. We will return later to this point.

The preferences of the federal government are a mirror image of the Lubicons'. The government's first choice is a treaty settlement, which it acknowledges itself obliged to make under the law. Its last choice is an aboriginal rights settlement, which by accepting the Lubicons' novel theory of adhesion might have repercussions on other pending claims and perhaps even on existing treaty obligations. It prefers the status quo of no settlement at all to recognizing the Lubicons' aboriginal rights claim; but, unlike the Lubicons, it might be prepared to accept the status quo forever because its only loss in doing so is some bad publicity from time to time.

Let us model the conflict between Canada and the Lubicons using a two-person, binary-choice game. Each player has two strategies: to offer an aboriginal rights (AR) or a treaty entitlement (TE) settlement. The parties only have to adopt the same strategies to achieve a settlement, as shown in Table 3.2.

Since the players are allowed to communicate with each other in this game, it would be a simple matter to coordinate their strategies and

48   Game Theory and Canadian Politics

TABLE 3.2
Model of Lubicon–Canada Game

		Canada	
		AR	TE
Lubicons	AR	AR settlement	No deal
	TE	No deal	TE settlement

Lubicons' preference order: AR settlement > No deal > TE settlement
Canada's preference order: TE settlement > No deal > AR settlement

achieve a settlement, except that their preference rankings dictate otherwise, as we can see in Table 3.3, where we use their preferences as ordinal payoffs in the game matrix (we follow the convention of reversing the order when we assign numbers to represent ranks; i.e., 3 = most preferred option, 2 = next, 1 = least preferred).

Given these payoffs, each player has a dominant strategy. The Lubicons should offer AR because they get a higher payoff no matter what Canada does. If Canada offers AR, then the Lubicons get their first choice (AR), which is better than their second, no deal (ND); and if Canada offers TE, the Lubicons get ND, which is better in their eyes than TE. For similar reasons, Canada has a dominant strategy of TE. If the Lubicons also play TE, the result is TE, which is Canada's first choice; and if the Lubicons play AR, the result is ND, Canada's second choice.

The game thus has a Nash equilibrium, shown in boldface in the upper-right cell of Table 3.3: ND, which is the second choice of both players. This is a Nash equilibrium because neither player has an incentive to make a unilateral move. Movement for either player without

TABLE 3.3
Lubicon–Canada Game with Ordinal Payoffs

		Canada	
		AR	TE
Lubicons	AR	3, 1	**2, 2**
	TE	2, 2	1, 3

movement by the other will result in the moving player dropping from second to third choice of payoff. In nontechnical language, there is a stalemate because each side prefers to continue without an agreement rather than to accept the other's first option, since the other's best choice is its own worst choice. This is the game known in the literature as Deadlock.[8]

Deadlock or stalemate ensues because neither party will accept the other's preferred principle of cooperation. If principle were not so important to both players, they ought to be able to reach a compromise by bargaining, 'splitting the difference' in agreed-upon proportions. But principle is important; that is precisely what makes the conflict so intractable. Hence, a break in the stalemate will require a change in the payoff matrix, which means a change in the way in which either or both players rank the outcomes. Let us look at the possibilities for the players to induce such changes in their opponents' preferences.

The Lubicons' political strategy for ending the conflict has long been to create embarrassment for the federal government, thus inducing the latter to reverse its ranking of TE and ND. A variant of the strategy is to embarrass or impose costs upon the Alberta government, hoping that Alberta will persuade Ottawa to change its preference order. Success in either version of this approach would produce a federal preference ranking of $AR > TE > ND$, resulting in the payoffs shown in Table 3.4.

In this version of the game, Canada no longer has a dominant strategy. Canada should play AR if the Lubicons play AR, and TE if the Lubicons play TE. Canada, of course, prefers the Lubicons to play TE, but will accept AR rather than miss an agreement altogether. The Lubicons, for their part, still have AR as a dominant strategy. Knowing that the Lubicons will play AR, Canada will do likewise, and the game

TABLE 3.4
Lubicon–Canada Game after Lubicon Political Success

		Canada	
		AR	TE
Lubicons	AR	**3, 2**	2, 1
	TE	2, 1	1, 3

Lubicons' preference order: AR settlement > No deal > TE settlement
Canada's preference order: TE settlement > AR settlement > No deal

will have a Nash equilibrium in the upper-left cell, representing an aboriginal rights settlement.

The Lubicons' political strategy is formally rational, but the evidence suggests that it is not empirically feasible. Until the band actually has a reserve, it has little leverage over the federal government. Most of the things it has done in the past, such as interfering with petroleum or forestry development, affect Alberta more than Canada; and Canada has not shown itself willing to budge because of discomfort for Alberta. The Lubicons perceived an opportunity to embarrass Canada in the 1988 Winter Olympics and did their best to exploit it, but they still did not succeed in changing Canada's preference structure. Nor has their appeal to the United Nations Human Rights Committee had any great impact.

The Lubicons could also resort to testing their aboriginal rights theory in court, and a legal victory would decisively change the federal government's preference order. If the courts held that the Lubicons still had unextinguished aboriginal title, the rule of law would compel Canada to make AR its first priority ($AR > TE > ND$), yielding Table 3.5.

Once again, the game has a Nash equilibrium. AR remains a dominant strategy for the Lubicons. Knowing that, Canada does better by offering AR, which is also its first preference. Both sides get the outcome they value most highly.

But the Lubicons have been reluctant to pursue their legal option, perhaps because they realize their theory is novel and has a high chance of rejection in court. Even worse, a judicial defeat for the AR principle would decisively undermine its political credibility. The political strategy of demanding AR does not require that the legal theory be demonstrably valid, only that it not be demonstrably invalid. The Lubicons thus have much to lose by going to court.

TABLE 3.5
Lubicon–Canada Game after Lubicon Court Victory

		Canada	
		AR	TE
Lubicons	AR	3, 3	2, 1
	TE	2, 1	1, 2

Lubicons' preference order: AR settlement > No deal > TE settlement
Canada's preference order: AR settlement > TE settlement > No deal

To formalize the reasoning, assume that the advantage of *AR* over *TE* is about $100 million (this figure is not exact, but is probably in the right order of magnitude). Assume further that the Lubicons' costs of obtaining *AR* can be ignored at this point. Whether *AR* is obtained by litigation or political negotiation, there will be lawyers' fees of several million dollars; but these will in the end be paid from some government account, since the Lubicons have few resources of their own. The expected value of undertaking a court action for *AR* is about $100 million multiplied by the probability of winning the action, or, if EV_L is the expected utility of litigation, and p_L is the probability of winning,

(1) $EV_L = 100 p_L$

Similarly, the expected value of getting *AR* by means of negotiations is $100 million multiplied by the probability of success in the negotiations (p_N), or

(2) $EV_N = 100 p_N$

I ignore here the possibility that the Lubicon might accept a negotiated settlement for less than $100 million because to do that would require them to bargain down their aboriginal rights principle, which they have been unwilling to do thus far.

These simple equations neglect the effect of unsuccessful litigation upon negotiation. In fact, the expected utility of litigation must be reduced by the effect that failure in litigation would have upon the prospects of negotiations. If the probability of success in litigation is p_L, the probability of failure is $1 - p_L$; so the impact of litigation failure upon the value of a negotiated settlement is to reduce it by $100 p_N (1 - p_L)$. Overall,

(3) $EV_L = 100 p_L - 100 p_N (1 - p_L)$

It will be rational to pursue litigation if the expected value of litigation as expressed in equation 3 is larger than the expected value of negotiation as expressed in equation 2, that is, if

(4) $100 p_L - 100 p_N (1 - p_L) > 100 p_N$

Dividing both sides by 100 and rearranging terms,

52   Game Theory and Canadian Politics

(5) $p_L - p_N + p_L p_N > p_N$ or
$p_L(1 + p_N) > 2p_N$

yielding

(6) $p_L > \dfrac{2p_N}{1 + p_N}$

Inequality (6) represents the condition under which it will be rational for the Lubicons to prefer litigation to negotiation as a way of pursuing AR. There is no unique solution, because p_L and p_N are independent of each other. But for any value of p_N, it is easy to compute the minimum value of p_L required to justify litigation. For example, if p_N is $1/3$, the threshold value of p_N is $2(^1/_3)/(1+^1/_3) = (^2/_3)/(^4/_3) = 1/2$. For realistic values of p_N, the conclusion is always the same: the Lubicons will resort to litigation only if they are appreciably more certain of its success than they are of success in negotiation. In case of doubt, prefer the political strategy because failure there does not pre-empt the legal strategy, whereas failure in litigation decisively undercuts the possibility of achieving AR through negotiations.

What are the options of the federal government for affecting the preferences of the Lubicons? It could try to make them forget about AR by winning a conclusive battle in court, which seems to have been the federal strategy in 1988 when it tried to initiate a lawsuit. But this strategy proved difficult in two respects. First, it was necessary to sue Alberta as well as the Lubicons, thus poisoning intergovernmental relations. Second, the Lubicons could not be compelled to cooperate in the legal process. It might have been possible to win a judgment even after the Lubicons refused to appear in court, but such a result might have made the federal government look like a bully and backfired in the wider realm of public opinion. Thus it is not surprising that Canada decided not to proceed with its litigation.

A more realistic strategy is to induce the Lubicons to prefer TE to ND; that is, to reorder their preferences from AR > ND > TE to AR > TE > ND. The reason that the Lubicons now prefer ND to TE is probably a belief that ND is temporary, and that they have the option to accept TE whenever they choose. They are correct in this belief to the extent that the federal government has repeatedly acknowledged that the Lubicons have a right to a reserve under Treaty Eight. Should the government try to repudiate this obligation, the Lubicons could almost certainly win a court judgment in their favour.

Stalemate at Lubicon Lake 53

Ottawa, however, is now undermining the Lubicons' position by recognizing and dealing with new bands who bleed off members from the Lubicons. Conclusion of an agreement with the Woodland Cree and other such bands in the future may reduce the Lubicons' numbers to the point where they could no longer justify the offer that is still on the table. Over time, the federal policy threatens to replace the 95-square-mile reserve (call it TE_1) with a smaller one (TE_2), probably accompanied by correspondingly smaller financial benefits. TE_1 remains on the table for now, but with an undefined time limit attached. A settlement will result if the Lubicons react to the implicit threat by putting TE_1 ahead of ND, perhaps in this ordering: $AR > TE_1 > ND > TE_2$. If we assume the government's new ordering to be $TE_2 > TE_1 > ND > AR$, Table 3.6 will result.

The Lubicons no longer have a dominant strategy, but they do have a weakly dominated strategy, namely TE_2, which gives results that are never better than those of either AR or TE_1, and sometimes worse. The Lubicons, therefore, will not play TE_2 because they can always do at least as well by playing something else. Canada, realizing that the Lubicons will not play TE_2, now has a weakly dominant strategy: TE_1. Knowing that, the Lubicons will achieve their best result by also playing TE_1, and an agreement will be reached in the central cell of the matrix. The reader can easily check that this is a Nash equilibrium because neither player can benefit through a unilateral move.

It is important to be clear about what this type of analysis accomplishes. We are only modelling a single instance, not testing a general model against a batch of empirical data. Hence, we are not able to establish a lawlike empirical statement describing the behaviour of many cases. In this respect, the analysis is quite different from the test of the lefty-righty baseball model in the preceding chapter. The value of mod-

TABLE 3.6
Lubicon–Canada Game with Threat of Smaller TE

		Canada		
		AR	TE_1	TE_2
	AR	4, 1	2, 2	2, 2
Lubicons	TE_1	2, 2	**3, 3**	2, 2
	TE_2	2, 2	2, 2	1, 4

Lubicons' preference order: $AR > TE_1 > ND > TE_2$
Canada's preference order: $TE_2 > TE_1 > ND > AR$

elling a single case of conflict is to clarify the thinking of the two sides. By looking at past behaviour, we can make plausible assumptions about preference orderings; and then we can say what future behaviour would be compatible with those preferences. We can also test behaviour for logical consistency, to see whether the actions of the government and the Lubicons at different times have been rational in the sense of moving the players as far as possible up their scale of values. By that test, it seems that both actors have been highly rational over long periods.

However, game theory is not prophecy. It cannot tell us whether the Lubicons will adjust their priorities as the government seems to hope, or whether they will think the government is bluffing. It cannot predict whether the federal government might change its priorities, perhaps in response to an election or other political pressures. But it does show us that both sides in this intractable dispute are behaving rationally by trying to obtain their objectives as they define them. In particular, Canada is now pursuing a realistic strategy that could induce the Lubicons to settle some day.

But if politics is an art, it is not like sculpture or painting, where one side acts and the other is passive raw material. It is more like drama without a script, or dance without choreography – in other words, improvisation. Either player may surprise us again with a new move.

4

Models of Metrication

The history of metrication in Canada furnishes excellent examples of the difficulties of implementing public policy. It is tempting for policy makers to think of themselves as artists painting on an empty canvas or sculptors working with clay, but policy implementation is more like a contest of forces. Policy initiatives are always met by popular reactions, and the final result is often different – sometimes wildly different – from what the policy makers intended. Game theory can be of considerable help in modelling what happens in such processes.

Throughout the nineteenth and twentieth centuries, the countries of the world have gradually switched from their traditional systems of weights and measures to the metric system. The United Kingdom began its transition in 1965, and in 1968 the United States Congress authorized a large-scale study of metrication. Hence, it is not surprising that in Canada the government of Pierre Trudeau, also elected in 1968, decided to go metric.

The first step was the *White Paper on Metric Conversion*, released in January 1970.[1] The necessary legislation was passed in 1971 with no real opposition in Parliament, and a preparatory commission, later renamed the Metric Commission, was appointed to plan the transition. That body coordinated the labours of over a hundred sectoral committees representing all areas of Canadian economic and social life, and integrated the results into a national plan published in 1974.[2]

The first step in the actual transition occurred on 1 April 1975, when the government weather service started to give the temperature in Celsius rather than Fahrenheit. In September of the same year, precipitation figures were converted from inches to millimetres and centimetres. In September 1977, road signs around the country changed from miles to kilometres, and in January 1979 gasoline stations started dispensing in litres rather than gallons. By December 1980, fabric had to be sold in metres rather than feet and yards, and December 1983 was the final cutoff for converting scales in retail grocery stores.[3]

At various points along the way, Canadian manufacturers started relabelling their products. Some simply made a 'soft conversion'; that is, they left their product in its original package or dimensions but made an arithmetic conversion to metric. You can still buy a 19-ounce can of tomatoes, but the label will now say 540 grams as well as 19 ounces. Other manufacturers went all the way to 'hard conversion,' changing the product to a 'rational' metric size. Thus motor oil and soft drinks are now sold in litre bottles, rather than in quarts carrying a metric equivalent on the label.

The intention was to make a complete transition to metric, but because of the integration of the Canadian with the American economy, attaining that goal has always depended on the willingness of the United States to go metric. Initially, that seemed likely to happen; in 1971 the secretary of commerce recommended that Congress adopt the metric system.[4] But political resistance developed, and Congress never moved forward.

Largely because of these U.S. developments, Canadians must now deal with two systems of measurement. Casual empiricism – talking to people, reading newspapers, visiting stores – reveals the following examples of measurement dominance.

Functional metric dominance with rational conversion:
- weather (temperature and precipitation)
- automobile travel (road signs, maps, gasoline, and motor oil)
- beverages (soft drinks, bottled water, alcoholic beverages)
- Olympic sports (track and field, swimming, cross-country skiing)
- science, medicine, pharmaceuticals
- fabrics and carpets

Continuing imperial dominance, with or without soft conversion:
- professional sports (baseball, football, basketball, hockey, boxing, wrestling, horse racing, rodeo)
- rural land description (the survey system of townships and sections would be tremendously expensive to change)
- body image (most adults know their height in feet and inches and their weight in pounds)

Models of Metrication   57

- personal clothing sizes (for example, 34-inch bra, 36-inch waist)
- home cooking (newspaper, magazines, and cookbooks generally use teaspoons and cups in their recipes; metric equivalents for these measures are not uniformly given)
- construction (precut lumber, nails, and screws are still in traditional sizes)

Apart from these fairly clear situations, consumers must be ready for almost any combination of metric and imperial measures. In a typical supermarket, most products will have metric information on the label, but some will be in rational metric sizes, others in imperial sizes with soft conversion. Unit prices will be posted in a confusing welter of dollars per kilogram, cents per gram, dollars per pound, cents per ounce, and so on. I could not find any research on how many customers can actually cope with all this code-switching, and how many just give up and buy what looks good. One finds a similar proliferation of weights and measures in a typical building-supply store, except that a great number of products imported from the United States have no metric equivalents at all on the label. The consumer will be looking at 26-inch lamps, 50-foot garden hoses, and two-gallon fertilizer sprayers. But one must also know metric to deal with litre bottles of antifreeze and 5-kilogram bags of grass seed, labelled without imperial equivalents.

This jumble of systems leads to amusing situations. For example, I recently asked for 200 grams of Colombian coffee beans in a specialty store: I phrased my request in metric because the stock was priced in metric. But the clerk asked if I would take 225 grams; 'that's how we sell it,' she said. Of course, 225 grams is a rounded-off version of half a pound ($453.6/2 = 226.8$). So even where metric conversion appears complete, the imperial system may lurk in the background.

Another oddity I have noticed concerns hiking and climbing in the Rocky Mountains. For mountain travel, it is important to know not only horizontal distance travelled but also elevation gain or loss. Trail signs for horizontal distance are now regularly posted in kilometres, and most hikers of my acquaintance seem to have switched to metric in this respect. However, perhaps because many topographical maps on which one reads elevations were published years ago in imperial and are still in use, hikers, even those who have immigrated to Canada from metric countries, almost invariably refer to elevation in feet. In the same sentence, without feeling any incongruity, a hiker will describe a trip as, say, being 16 kilometres long with 3500 feet elevation gain.

Metric is taught systematically in the public schools, whereas, at least where I live, the imperial system is taught only briefly, once in junior high school. My impression is that young people are growing up with a reasonable command of metric but without an overall grasp of imperial. They become acquainted with a few units to cope with particular situations – inches to buy clothes, pounds for their own weight, yards if they are football fans, teaspoons and cups if they like to cook, acres if they live on a farm – but they are unclear about how these units relate to one another or how to convert them to metric.

Metric lives as a system, while imperial survives in disjointed fragments; but this situation does not mean that imperial will soon fade away. Because of the North American Free Trade Agreement, our economy is more than ever integrated with that of the United States. For the foreseeable future, Canada will have to be a functionally bimetric, as well as bilingual, country.

Coordination Games

Weights and measures belong to the class of human contrivances known as conventions; other well-known examples are language, money, and table manners.[5] The utility of conventions lies in their general acceptance, not in any inherent value. English is not intrinsically superior to French, and French is not superior to English, nor is either superior to Swahili. Any idea can be expressed in any language, but it is of little use to speak a language unless the people around you also speak it. The same is true of money. The yen is just as good as the dollar, but a fistful of yen is not as useful in Toronto as a fistful of dollars because most stores accept only dollars. Similarly, the most important thing about weights and measures is not *what* they are but *that* they are generally understood and accepted.

The underlying model of convention is known as a *coordination game*, of which the simplest version is a two-person ordinal game. Imagine that you are walking down a narrow corridor when someone else approaches. If you step to your right and the other person steps to his right, you will pass each other unhindered. The same is true if you both step to your respective left. Score an unimpeded passage 1 for each of you. But if you step right and he steps left, or vice versa, you will bump into each other and waste time. Score that outcome 0 for each of you. The precise numbers don't matter here; they are merely indicators of superiority (1 is preferred to 0). The situation can be summarized in the ordinal game of Table 4.1.

TABLE 4.1
Coordination Game

| | | Person B | |
		Right	Left
Person A	Right	**1, 1**	0, 0
	Left	0, 0	**1, 1**

There are two Nash equilibria in this game, with payoffs of (1, 1), corresponding to the two strategy pairs of right-right and left-left; and both equilibria are equivalent for both players. In simple language, you are equally happy to be at right-right or left-left; but you definitely want to be at one of those points, rather than at right-left or left-right. Moreover, if you are at one of the equilibrium points, say right-right, you would not try to switch to the other, left-left, except in coordination with the other player. A unilateral move away from a Nash equilibrium point can only make you worse off.

Let us now explore the question of how a convention might arise over time.[6] To make the model more useful for this purpose, imagine a population of players who interact with each other repeatedly, one-on-one, in different pairwise combinations – say the inhabitants of a village who drive their wagons through a central intersection. They are frequently playing the two-person coordination game with a variety of partners. In the beginning, if there is no rule about left or right, you might as well play randomly, going right half the time and left half the time in no particular order. If all were playing in this way, you would get 1 half the time and 0 half the time, for an average payoff of 0.5.

Note, however, that this is a statistical average over a long period. In any shorter period, owing to random fluctuation, there might be more players going right (or left). Or one player might see that, if everyone else is playing randomly, it really doesn't matter what she does, so she might decide to go always right (or left). Either development, or both together, would tilt the results over some period, so that instead of a 50–50 split, there is, say, a 55–45 split in favour of right (or left). Once other players realize the existence of this disparity, no matter what its cause, it would be rational for them to start playing right (or left) all the time in order to take advantage of it. If on average 55 per cent of other players' moves are to the right (or left), you will get 1 on 55 per cent of occasions if you always move right (or left). This choice would be self-reinforcing because, as more players choose to move always to one side, the more logical it would be for the remaining players to move the same way.

Thus we would expect a convention to arise over time even in the absence of central authority or conscious agreement. Simply through repetition, players would settle on one of the two possible equilibria. However, we would not expect the same equilibrium to emerge everywhere; it should be left-left in some places and right-right in others. Broadly speaking, this pattern seems to correspond with the actual history of rules of the road around the world. Some localities settled on right-right, others on left-left.

A similar pattern is also discernible in the field of weights and measures. Innumerable local communities came to agreement on weights and measures without any exercise of authority. Parts of the human body were at first taken as models, like the hand or foot, then human activities, such as the area that a man could plough in a day.[7] Such measures were accepted because they were useful, and they were useful because they were accepted.

Yet the historical record also shows an involvement of government in weights and measures from the dawn of recorded history. 'The right to determine measures,' says the leading historian of the subject, 'is an attribute of authority in all advanced societies.'[8] One reason for governmental involvement in the field is easily seen in the logic of the coordination game. Since a Nash equilibrium is self-enforcing, it emerges spontaneously and is difficult to change thereafter. Once players have adopted a standard, no one will want to change unless everyone else is also changing at the same time. Yet as political regimes expanded in size, they came to embrace many communities with their own local weights and measures. In other words, with the growing size of political communities, self-generating conventions inevitably come into contact and conflict with each other.

One can envision an evolutionary process in which different systems of weights and measures compete with each other, giving rise in the end to the victory of one, or perhaps to a merger of several systems; but this result would necessarily take a substantial period during which much effort would be wasted on comparison and translation of measures. Hence it would be tempting for a well-meaning sovereign, concerned about economic efficiency, to solve the coordination game by decree. Government could make an authoritative announcement to signal everyone to change in unison, thus restoring the benefit of a universally accepted set of standards.

Models of Metrication   61

Let us take this pleasant little story at face value, even though the historical record suggests that sovereigns had stronger and less benevolent motives for intervening in measurement. By asserting their control over measures, sovereigns could alter the contents in ways beneficial to themselves; for example, by redefining the units of grain in which tribute had to be paid. Prior to the democratic age, political struggles over weights and measures had more to do with exploitation by authorities than with the advantages of coordination. Nonetheless, the need for coordination is real enough.

The potential benefit of government involvement was ratcheted upward with the design of the metric system during the French Revolution. In comparison with all traditional systems, including the imperial one, the metric system has three intrinsic advantages that justify a claim of inherent superiority. First, because metric units are defined in terms of invariant physical features of the universe, they are protected against exploitive manipulation by sovereigns. Second, the units of length, area, mass, volume, and so on, are directly related to each other. Third, all units are defined in terms of the decimal system, making calculations easier, at least for those who have mastered the decimal system.

There is something to be said against these alleged advantages, but for the moment let us accept that metric is a superior system of measurement. For people who already have the imperial system of weights and measures and are considering metric, a useful model is the *assurance game*, portrayed in its simplest two-person form in Table 4.2.

If metric is a superior system, it is better to use it than to use imperial, but only if the other person is also using metric. Two people using different systems produce only confusion when they deal with each other. The assurance game would not be interesting if we were starting from scratch because each party would independently choose metric in a win-win situation. The game starts to become interesting if we stipu-

TABLE 4.2
Assurance Game

		Person B	
		Imperial	Metric
Person A	Imperial	**1, 1**	0, 0
	Metric	0, 0	**2, 2**

late that metric is a new invention confronting players who are already using imperial.

Of course, (2, 2) is obviously superior to (1, 1) for both players. In the jargon of game theory borrowed from economics, (2, 2) is Pareto-optimal; that is, it makes both players jointly as well off as they can be within the constraints of the game. However, both (1, 1) and (2, 2) are Nash equilibrium points; if you move unilaterally from either one, you make yourself worse off. To justify shifting from imperial to metric, you have to have *assurance* that the other player will do likewise. Hence (1, 1) is a *trap;* that is, a Pareto-inferior Nash equilibrium. Political authority can be used to supply the assurance that both will move, thus allowing each to exit the trap and improve his payoff, both individually and jointly.

The two-person model illustrates the nature of the problem, but an extension is needed to represent the complexity of a larger society. The graphic approach displayed in Figure 4.1, originally developed by the economist Thomas Schelling, is frequently used for this purpose.[9]

In this application, the depiction captures the fact that the two-person assurance game is repeated over and over in the transactions of daily life. Each customer in a store, each reader of a newspaper, each listener to a radio weather forecast is engaged in a two-person game; but the social reality is the sum of these games. In that respect, it is like an n-person game, but without coalition formation. The players are not trying to win by forming exclusive coalitions; they are merely trying to get through a long series of transactions as efficiently as possible. Some authors call this sort of situation a compound game.[10] We can also think of it as a non-cooperative n-person game, in which many people are involved but do not make conscious agreements with each other. No matter what we call it, it represents an interesting transition from two-person to n-person games. Clearly, the underlying model of the interaction is two-person; but, just as clearly, many actors are involved.

Figure 4.1 is a two-dimensional space. The horizontal axis represents the variable p, which is the percentage of people who use the metric system, ranging from 0 per cent at the left edge to 100 per cent at the right edge.[11] The vertical axis represents the utility derived from using either the metric or the imperial system, ranging from positive (advantageous) above the horizontal axis to negative (disadvantageous) below the horizontal axis.

The two solid diagonal lines represent utility functions for imperial and metric on the assumption that the two systems have the same

FIGURE 4.1

Depiction of Metrication Decisions with Schelling Curves

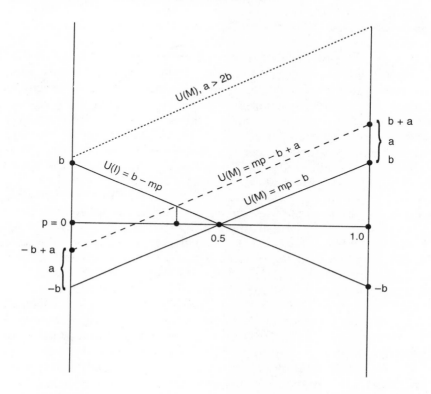

inherent efficiency and that the players have no preference for either. Under these conditions, the value of using either is totally dependent upon what the other players are doing. If everyone is using imperial ($p = 0$), there is a positive benefit, b, derived from using the system; and there is a similar loss, $-b$, derived from trying to use metric in an imperial world. But for every person who decides to use metric, an amount, m, is subtracted from the value of using imperial and added to the value of using metric. If everyone is using metric ($p = 1.0$), the value of doing so is b, and the cost of trying to stick with imperial is now $-b$.

The two lines intersect each other at the same point where they cross the horizontal axis; at this point, $p = 0.5$. As long as the imperial line is above the metric line, using imperial is a dominant strategy for a rational player; but where the metric line is higher, metric becomes the dominant strategy. In simple language, this means that, if the two systems are equally efficient and you are equally proficient with both, you should use whichever one is being used by more than half the population.

The point where lines cross each other is usually called an equilibrium, and this crossover is an equilibrium in the sense of representing the point where the benefits of using metric and imperial are exactly in balance. At $p = 0.5$, the choice of measurement system is a matter of indifference. However, the equilibrium is unstable (self-destructive) rather than stable (self-correcting). At $p = 0.51$, using metric is more advantageous than using imperial, and metric becomes a dominant strategy. Even the slightest departure in either direction from $p = 0.5$ creates a dominant strategy and a new, stable equilibrium of all-metric ($p = 1.0$) or all-imperial ($p = 0$). These Nash equilibria are the only two stable equilibrium points; the mixed-strategy equilibrium is not sustainable.

The broken line represents the utility function for metric on the assumption that metric has a net advantage, a, over imperial. We have already mentioned three intrinsic advantages of metric, which are real enough. However, imperial also has certain advantages that have to be taken into account.[12] First, many imperial units are derived from parts of the human body. A foot is about the length of a grown man's foot, an inch is a little less than the last joint of the thumb, and two yards is about the span embraced by a man extending his arms at right angles to his body. This corporeal origin of at least some imperial units gives them an intuitive appeal and, in a pinch, facilitates estimation (pacing off a field, for example). Second, since imperial units evolved in daily use, their size approximates the requirements of many practical situations. The foot, the pound, and the pint are all close to the size of many things used in daily life. In contrast, the metre, the kilogram, and the litre are all rather large, and the centimetre, the gram, and the millilitre rather small, for many practical uses. Third, the decimal system, which requires a zero, is a relatively late development in arithmetic, and people without a lot of formal education find it mysterious. In daily life, people find it quite convenient to multiply and divide quantities by small integers, particularly 2, 3, and 4. Thus, the division of the foot into 12 inches is handy because 12 can be divided by 2, 3, 4, and 6 without a remainder. Similar integral ratios unite the teaspoon, tablespoon, cup, pint, quart, and gallon to each other. Another example is the partition of the inch through successive division by 2: $1/2, 1/4, 1/8, 1/16, 1/32$.

Probably dwarfing all these factors in significance is the simple fact that society has a large capital investment in imperial as the established system. This investment has many aspects: the learning of the system by individuals; hand and machine tools calibrated in imperial; machines that, because they were manufactured according to imperial engineering standards, need to be repaired with imperial parts; craftsmanship dependent upon imperial measures, such as the ability to build frame housing using 2" x 4" studs and 4' x 8' sheets of plywood; and entrenchment in the legal system, such as the rectangular survey, which locates land in terms of townships, sections, and chains. All this mental and physical capital must be replaced to effect a thoroughgoing conversion to metric.

None of these factors can be precisely quantified. However, I am prepared to believe that, at least in the modern world where formal education is widespread, metric has a net advantage over imperial and other traditional systems of measurement. The decisive evidence is that, since its invention, metric has been gradually adopted throughout the world, and that today the United States is the only major nation that has not made a formal commitment to it. Moreover, no nation, once having made a transition to metric, has ever gone back to its prior system. In addition, metric has acquired a further advantage as it has been adopted throughout the world. Whatever its merits within a single society, metric is now the world standard, an important factor in an age when international trade and travel are becoming more important every year.

For all these reasons, it seems sensible to grant the advantage of metric and to write the equation $U(M) = a - b + mp$, which means in words that the utility of using metric in an imperial society is equal to the metric advantage (a) minus the cost of going against the imperial standard ($-b$) plus a constant (m) times the proportion of people who are also using metric (p). The broken line in Figure 4.1 represents this utility function incorporating the metric advantage. Note that this line lies above the solid line for metric. Therefore, the crossover point between the imperial line and this second metric line is to the left of the point where $p = 0.5$. The greater a is, the further the crossover point will be displaced to the left. At the extreme, when $a > 2b$, the metric line will lie above the imperial line even when $p = 0$, which means the metric will always dominate imperial as a strategy, and we could expect spontaneous adoption of metric within the society. This situation is represented by the dotted line in Figure 4.1.

From the fact that, except in the fields of science and medicine, metric adoption has not been spontaneous and has always required some degree of government action, we can conclude that $a < 2b$, which means that the broken line is the more realistic depiction. Metric will become a dominant strategy only if a certain threshold of use is reached. If a is large, that crossover point might be quite a bit less than 50 per cent, but it will still be greater than 0. So the question remains: how will the initial users of metric be motivated so that the crossover point can be reached? Once we get to the crossover, metric becomes the dominant strategy and conversion proceeds spontaneously. But how do we get to the crossover?

Of course, this is not much of a problem for an authoritarian government. In a society where individual rights and consent of the governed don't matter, the government can reduce the value of b by laying heavy penalties on the use of the traditional system of measures. Merchants, manufacturers, and consumers can be fined or even sent to jail for dealing in the traditional system. The police can destroy scales and other devices calibrated in the old system. In fact, steps of this type were finally taken in France in 1840 to complete the conversion to metric that had been ongoing for 45 years. As Kula writes, 'The reform that standardized weights and measures, which had been so ardently desired for centuries and so widely demanded by the common people on the eve of the Revolution, extolled by so many of the truest revolutionaries and conceived by the finest scientific minds of the day, had, ultimately, to be imposed upon the people.'[13]

Modern Canada, in contrast, is a liberal democracy in which little can be done by simple decree. The government's White Paper emphasized leadership and consultation, rather than coercion, and stated explicitly: 'No legislative action is contemplated which would make mandatory a general use of metric in place of inch-pound units, although some legislation may prove desirable to foster familiarity with metric units.'[14] The amendments to the *Weights and Measures Act* passed in 1971 gave legal status to both metric and imperial measures.[15] Cabinet received the power to require that measuring devices used in trade be capable of giving metric readings,[16] but that is a far cry from the power to outlaw the use of imperial measures among the general public.

In fact, the federal government, in collaboration with the provincial governments, has relied chiefly upon policies of persuasion, leadership, and conversion within its own sphere. Typical policies include:

- teaching more metric and less imperial in school,
- switching to metric in the reports released by the federal government's weather service,
- converting signage on public highways to metric,
- using sectoral committees to encourage industries to metricate, and
- requiring that metric scales and metres be used in retail trade

The impact of this ensemble of policies is to increase a, thus bringing the metric and imperial lines closer together, shifting the crossover point further to the left, and reducing the value of p at which metric becomes a dominant strategy. Yet a problem remains. Even if governmental measures succeed in reducing p from 0.5 to, say, 0.1, that still means 10 per cent of a large society must make the leap before metric becomes a dominant strategy. What is to motivate the initial switchers?

One hypothesis starts from the fact that the broken line represents a standardized or average utility function for people in general. In reality, however, there are significant differences among individuals. Recent immigrants from metric countries probably already have a preference for metric. Native-born Canadians with a lot of formal education, particularly in the natural sciences, are also familiar with metric and may very well prefer it. In other words, there are people whose individual metric advantage is larger than the average a. This means that the broken line is really the central tendency of a broad band of people, half of whom lie above the line. As government policy brings the metric line closer to the imperial line, some of those with a larger metric advantage may actually be pushed above the imperial line altogether, so that metric becomes a dominant strategy for them. If these individuals make the conversion, they may raise the value of p to the point where conversion becomes worthwhile for others whose metric advantage is not so high, and so on.

To illustrate the principle, imagine that 11 hikers come to a narrow ledge that fewer than 10 per cent of the group would cross if they were making an individual decision. Suppose that the most reckless hiker in this group will cross virtually any ledge; and that each of the other 10 is in a different decile of recklessness, ranging from the top 10 per cent of the population to the bottom 10 per cent. That is, the second most reckless hiker of the 11 will cross a ledge that 10 per cent of people will cross, the third most reckless will cross a ledge that 20 per cent will cross, and so on. If the most reckless hiker will cross, the whole group will cross, because the passage of each additional one satisfies the need of the next one for progressively greater reassurance. The group as a whole is not reckless (the median member is only at the 50th percentile); but all will cross as long as they take their bearings from what others in the group are doing.

Schelling called this phenomenon *tipping* and discussed it in the context of residential segregation. Imagine a few blacks moving into a previously all-white neighbourhood. If there is an exodus by whites who are intolerant of even a small black presence, the black percentage will rise, causing further white departures in turn. Some whites who might have stayed in a 10 per cent black neighbourhood may start to feel uncomfortable if that percentage rises to 20 per cent or 30 per cent, and so on. 'The problem is that the 70:30 percent mix is not a stable equilibrium. If this mix is somehow disrupted, as chance is sure to do, there is a tendency to move toward one of the extremes.'[17] Broadly speaking, what the Canadian government tried to do was to tip, rather than force, the Canadian people toward the adoption of metric.

Interestingly, however, the tipping process seems to have worked in some domains better than in others. As pointed out earlier, Olympic sports, gasoline, and alcoholic beverages are now metric, while professional sports, clothing sizes, and construction lumber are still imperial and show no signs of undergoing conversion. This observation leads to one final modification in the model. Instead of representing the metric utility function by a single line, there should be a family of lines, as shown in Figure 4.2.

In Figure 4.2, the higher the line, the greater the metric advantage in that domain. In scientific and medical research, where the metric advantage is very great, metric seems to be a dominant strategy across the board; metrication took place spontaneously even before the government decided upon the metrication of Canada. The conversion of weather information, road signs, and liquids happened quickly and relatively painlessly, whereas conversion of manufactured products was slower and in many cases amounted to soft conversion at best. Finally, some areas of life, such as professional sports, have remained untouched by metrication.

Three generalizations suggest themselves about the hard-to-metricate domains of life:

- They involve large amounts of physical capital such as tools and machinery (manufactured goods, construction timber, records of land titles).

FIGURE 4.2
A Family of Schelling Curves

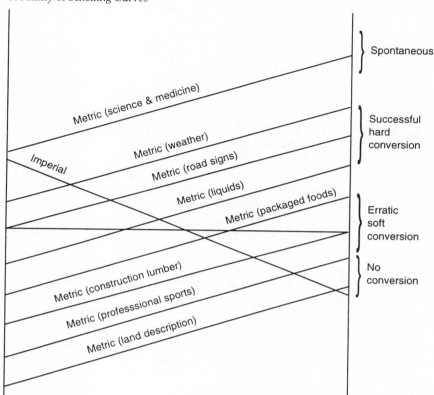

- They are tightly interwoven with the U.S. economy and society, where metrication has never taken place (professional sports, some manufactured goods).
- They involve body image, where there is a high emotional investment (personal height and weight, clothing size).

In contrast, the domains where metrication occurred either spontaneously or with just a little nudging could be characterized as follows:

- They are abstract, involving little physical capital (weather reports, road signs).
- They are connected to the world at large, not just the United States (Olympic sports, scientific and medical research).

70 Game Theory and Canadian Politics

- They involve bulk products not closely connected to our bodies (gasoline, wheat).

The presence or absence of these factors means that it is far more costly to metricate in some domains than in others. It stands to reason that, where the metric advantage is smaller, people will be more resistant to conversion and more coercive policies will be required.

The results of Canada's conversion to metric are rather unsatisfying. The advocates of conversion foresaw a simpler, more rational, more efficient world. Children would spend less time learning weights and measures in school, engineers would save time doing their calculations and make fewer errors, the arithmetic of daily life would become easier and simpler. But it hasn't turned out that way. Canadians now have to know two systems of measurement to cope with daily life. Some areas are purely metric or purely imperial; but in others, such as grocery and building-supply stores, the two systems are jumbled side by side.

The modelling exercises undertaken in this chapter help to reveal the inner logic of what happened. It was neither administrative incompetence nor lack of political will that got us where we are today. The well-meaning desire of our government to solve an assurance game without indulging in heavy-handed coercion, combined with the unexpected failure of the United States to carry through its own metric conversion, left us with our variegated set of outcomes, all of which are rational in context. Metrication has proceeded in those domains where it is desirable under present circumstances; other domains have remained imperial. Trying to force the pace will only bring expense, evasion, and noncompliance. If there is any larger lesson, it is that dreams of a simpler, more orderly world seldom come true.

Wider Horizons

Schelling's graphic approach to representing n-person games can be applied in many other areas of Canadian politics, even where the facts are quite different from the metrication example. For instance, there is concern about the reappearance of measles, whooping cough, and other childhood diseases for which effective vaccines have long been available.[18] There are complexities (some vaccines need to be administered more than once), but we can initially model the decision as a binary choice between vaccination or not. The net benefit of vaccination depends on the effectiveness of the vaccine in preventing illness (no vaccine is 100 per cent effective), the economic cost of the vaccine (trivial in most cases), and the medical risk (all vaccines produce occasional side effects, including serious illness and even death). But the equation must also include the percentage of *other people* who undergo vaccination.[19] If no one else is vaccinated, the benefits to me of undergoing vaccination are high, because I am protecting myself against the many vectors of contagion that surround me. But if almost everyone else is vaccinated, the relative benefit of being vaccinated is much less because the possibility of infection is already much reduced. In such circumstances, the risk of illness caused by the vaccine might well be greater than the risk of infection if I am not vaccinated.

The two-person model that provides microfoundations for this situation is known as Chicken. Each player's first preference would be for the other to be vaccinated, which would eliminate any possibility of infection for herself. But if the other is not vaccinated, a player would rather protect herself through vaccination than be exposed to possible contagion. The full set of preferences is shown in the ordinal payoff matrix of Table 4.3.

There are two Nash equilibria in this game, both lying on the off diagonal. They represent the two unequal outcomes, where one player is vaccinated and the other is not. Neither player benefits from a unilateral move away from one of these points. Note that, in contrast to the coordination and assurance games, there is an element of competition here. You don't win by doing the same thing as the other player; you win by doing something different and forcing her to take the less desirable outcome.

Having Chicken as microfoundations produces an interesting result in the *n*-person game, as shown in Figure 4.3.

TABLE 4.3
Chicken Game as a Model of the Vaccination Decision

		Person B	
		Vaccinate	Don't vaccinate
Person A	Vaccinate	2, 2	**1, 3**
	Don't vaccinate	**3, 1**	0, 0

FIGURE 4.3
Schelling Curve Model of Vaccination

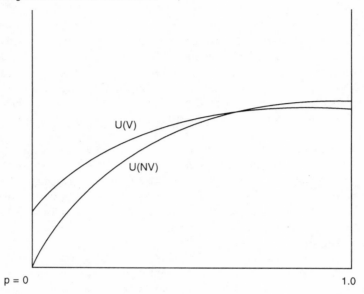

In Figure 4.3, the horizontal axis represents the percentage of people (p) who undergo the vaccination. The line $U(V)$, representing the utility of getting vaccinated, slopes upward because those who undergo vaccination confer spillover benefits on others. Each person who gets vaccinated reduces by a small amount the risk of infection for others by removing one person from the pool of possible carriers. That is, the protection I enjoy will be greater if, say, 75 per cent of people are vaccinated than if only 25 per cent are. However, the curve also tends to level off because there is a limit on the amount of immunity that can be acquired and conferred; protection is never perfect. The other line, $U(NV)$, represents the utility of not being vaccinated. It starts at zero but rises steadily as p increases. At some point the line $U(NV)$ will cross the line $U(V)$ because those who do not get vaccinated do not expose themselves to the risk and costs entailed by vaccination.

In this model the crossover point is a stable, self-maintaining equilibrium. If for some reason fewer people get vaccinated and p declines, we would move to the left, $U(V)$ would be above $U(NV)$, and vaccination would be a dominant strategy. But if more people get vaccinated, we would move to the right, $U(NV)$ would lie above $U(V)$, and Non-Vaccination would become dominant. In both cases, the change in behaviour that follows a change in p is in the opposite direction, and thus tends to move the system back to p. This kind of self-enforcing, stable equilibrium is like a mixed-strategy equilibrium, except that without actual cardinal measures we cannot calculate the equilibrium point as we did for baseball.

However, we can still make a qualitative prediction about the real world: it is difficult to achieve universal vaccination. It is not impossible, as shown by the eradication of smallpox, but it is difficult.[20] At some point of success in mass immunization, it becomes rational for those not yet vaccinated to avoid the procedure, because to go through with it carries more cost than benefit. Failure to be vaccinated is usually attributed to sloth, ignorance, or superstition; but this analysis suggests that, at a certain point, avoiding vaccination can be a rational exercise in pursuit of self-interest. It may be immoral, because the Non-Vaccinators are free-riding on the willingness of the Vaccinators to expose themselves to risk, but it is not irrational. Government can try to shift p as far as possible to the right by levying legal penalties for noncompliance, but evidence from this as well as other fields shows that even draconian penalties seldom achieve 100 per cent compliance when they try to prevent people from acting in a self-interested way. That does not mean we should not try, but it does mean we should be realistic in our expectations. Getting everyone vaccinated is never as easy as getting everyone to drive on the same side of the road or to talk about the weather in Celsius degrees.

5

How Many Are Too Many?
The Size of Coalitions

Coalition Theory

Two-person games, as employed up to this point, have real but limited application to politics. There are occasional head-to-head contests in politics, such as the Lubicon deadlock, but most political conflicts involve more than two players. Even contests that might be considered two-person games for certain purposes, such as the electoral struggle between the Democrats and Republicans in the United States, are in another perspective better seen as conflicts between two large segments of the population – that is, between coalitions.

The fundamental unit of political activity is the coalition, defined by one scholar as 'the joining of forces by two or more parties during a conflict of interest with other parties.'[1] Involving both conflict and cooperation, coalitions draw a boundary within which cooperation takes place to defeat or gain advantage over an external opponent. Political coalitions are formed precisely to exclude others and thereby exercise power over them.[2]

The study of politics, therefore, requires a theory not just of two-person but of n-person games. A good place to start is with the three-person game known as Divide the Dollar. Assume there is a payoff of one dollar, to be divided according to the decision of a majority of the players. The only rule is that a majority must form before the dollar can be divided. With three players, there are four possible types of coalitions: a grand coalition of all three players, a simple majority coalition of two, a minority coalition of one, and a null coalition of zero. (The null coalition is of no practical importance; it's just the theoretical complement to the grand coalition, as the minority coalition is the complement to the majority coalition.) The rule means that a grand coalition or a simple majority can divide the dollar, but a minority or null coalition cannot.

Given the general assumption of game theory that rational players want to have more rather than less, how would we expect them to play Divide the Dollar? It is easiest to think about the play as a series of moves, even though we are actually describing a train of thought in the minds of the players. They might begin by forming a grand coalition and dividing the dollar $\{1/3, 1/3, 1/3\}$. However, this *imputation*, as it is called in n-person game theory, would be dominated by any of the following three imputations: $\{1/2, 1/2, 0\}$, $\{1/2, 0, 1/2\}$, or $\{0, 1/2, 1/2\}$. In plain English, if two players decide to move from a grand coalition to a majority coalition, they can increase their payoff from $1/3$ to $1/2$ by excluding the third player and reducing his payoff from $1/3$ to 0. Game theory has no way to predict which majority coalition will form in these circumstances, so there is no exact equivalent of a minimax or Nash equilibrium; but there is a *solution set* in the sense that we would expect the formation of one of the three simple-majority coalitions.

A solution set is not stable in the way that a minimax solution or Nash equilibrium is stable. Suppose, for example, that players A and B form a coalition with the imputation $\{1/2, 1/2, 0\}$. Player C, who gets a payoff of 0 by virtue of being excluded, can break up that coalition by offering a different division of the dollar to A, say $\{0.51, 0, 0.49\}$. Although both A and C would benefit from that move, the new imputation would also not be stable because B could upset it by offering $\{0, 1/2, 1/2\}$ to C. At this point, we would be back in the solution set, although with a different imputation than the one we started with. The concept of a solution set allows for cycling outside itself, but also predicts a return to the solution set.[3]

Note that the simple-majority coalitions could also be called *minimum winning coalitions* (MWCs). In general, the solution set of Divide the Dollar and similar games will consist of minimum winning coalitions that allow the MWC members to take a maximum share of the payoff. With five players, one would expect the solution set to consist of the MWCs of size $n = 3$, with individual payoffs of $1/3$, rather than the larger-than-necessary winning coalitions of size $n = 4$, with individual payoffs of $1/4$.

The concept of minimum winning coalition gets more complicated in voting games if all the players do not have the same voting power. Suppose A and B both have two votes, whereas C has only one. The

76   Game Theory and Canadian Politics

coalition *AB* wins by a margin of 4–1, whereas *AC* and *BC* win 3–2. In such cases, is the *MWC* to be defined in terms of a minimal number of *players* or a minimal number of *votes*? We will take up this complication in the next chapter; in this chapter we will deal only with coalitions where the players have equal weight.

Another point to note is that we are now thinking about cooperative games. *N*-person situations came up before in the Schelling curves used to analyse metrication, but those were noncooperative models in which participants made individual decisions without attempting to reach agreements with other players. In a *cooperative game*, the rules allow players to make agreements with one another to pursue coordinated strategies. A coalition is a set of players who agree to pursue a common strategy.

The Size Principle

William Riker, in *The Theory of Political Coalitions* (1962), made the first major application of cooperative n-person game theory to political analysis. According to Riker:

In n-person, zero-sum games, where side-payments are permitted, where players are rational, and where they have perfect information, only minimum winning coalitions occur.

In social situations similar to *n*-person, zero-sum games with side-payments, participants create coalitions just as large as they believe will ensure winning and no larger.[4]

The zero-sum condition applies to games like Divide the Dollar because the payoffs of the winning coalition equal the losses of the other players. The side-payments condition can also be called transferable utility; it means that the total payoff can be divided and redivided among the players without constraint. This condition does not apply to many *n*-person games; for example, to voting games, where the payoff is an indivisible decision or a piece of legislation. The nine justices of the Supreme Court of Canada are playing a voting game when they decide a case, as are the 301 members of the House of Commons when they vote on a bill; but the payoff is an indivisible decision that cannot be distributed among the members of a winning coalition. Hence, we would not expect Riker's size principle to apply to such cases; but we would expect it to apply in cases such as formation of a cabinet to control the executive government, or formation of an international alliance to defend territory or fight a war. Cabinets have many divisible benefits (jobs, contracts, policies) to distribute to their followers; and states that participate in international alliances can share out conquered territory as well as apportion the expenses of fighting and defending against wars.

Riker's translation of the size principle from game theory to real-world situations relaxes the assumption of perfect information. He writes: 'The greater the degree of imperfection or incompleteness of information, the larger will be the coalitions that coalition-makers seek to form and the more frequently will winning coalitions actually formed be greater than minimum size.'[5] This assumption has a couple of practical implications. First, since information in the real world is never complete, coalition-builders will want to have some margin of safety. If the House of Commons has 301 seats, it would be foolhardy to aim to win only 151. Riker's analysis doesn't tell us exactly how many seats parties will try to win, but it certainly implies something above 50 per cent + 1, although much closer to that figure than to 301. Second, the margin of error will need to be larger in times of great uncertainty. One oft-noted example is the tendency of parties to form a grand coalition of national unity during major wars, when the survival of the state is threatened. Something like this happened in Canada during the First World War, when a large fraction of the Liberals joined the Conservatives in the Union government and ran on a common platform in the election of 1917. Riker's theory allows for such exceptionally large coalitions in unusual times, but also predicts that they will tend to break apart quickly once the situation stabilizes, as happened to the Union government after the end of the Great War.

Much of Riker's book consists of historical examples illustrating his theory. Perhaps the most persuasive are the three instances of total world war in the last two centuries, each of which gave rise to a near grand coalition, an overwhelming majority to defeat the aggressor. First was the alliance of Great Britain, Austria, Prussia, and Russia that finally defeated Napoleon. Second came the Allied powers of the First World War – Great Britain and the Dominions, France, Italy, Russia, Japan, the United States, and many smaller countries – against Germany, Austria-Hungary, Bulgaria, and the Ottoman Empire. Third came the Allied powers of the Second World War – Great Britain and the Dominions, France, the Soviet Union, the United States, China, and many smaller powers – against Germany, Italy, and Japan. Each episode illustrates the connection between uncertainty – in these cases, a threat to survival – and the formation of a larger-than-normal coalition.

Equally interesting is the fate of each grand coalition. The victors of the Napoleonic wars tried to establish the Concert of Europe, but that quickly foundered on imperial rivalries. The 20th-century coalition partners did not have much better luck with the League of Nations and the United Nations. The United States did not enter the League of Nations at the founding, even though it was the brainchild of U.S. President Woodrow Wilson; and Italy and Japan subsequently split off to join Germany in the Axis powers. The rivalry between the Soviet Union and the Western powers was evident even before the end of the Second World War, and by 1946 people were talking about the Cold War. Once the threat to survival was relaxed, the larger-than-normal coalition disintegrated.[6]

Note that Riker's size principle is by no means a self-evident truism. Politicians in democratic countries routinely speak as if they wanted to get every vote and win every seat. Anthony Downs explicitly assumed that 'every government seeks to maximize political support,'[7] just as entrepreneurs try to maximize profits. In contrast, Riker asserts that politicians aim for just enough support to take and keep control, with an adequate margin of error to allow for contingencies. Having too much support could be almost as bad as having too little because you have to reward your supporters to keep them in the coalition. The larger the coalition, the smaller the individual share of benefits that can be provided, and the greater the risk that a crucial portion of your members might be won over to some other coalition.

Political Parties

At the level of the parliamentary caucus, parties are coalitions in a strict sense. The caucus consists of a number of members of Parliament who deliberately cooperate to pursue common policies and who may be disciplined for breaking them, as in the expulsion of John Nunziata from the Liberal caucus in 1996 after he voted against the government's budget. In a somewhat looser sense, the party organization is also a coalition. Canadian federal parties vary in size, but they all have tens of thousands, sometimes over a hundred thousand, members who have paid a fee to join. Cooperation within the organization is real but loose. Many members, but not all, give time and money to support party activities; and some nominal members may in fact oppose the party on certain matters. Finally, the millions of supporters who vote for candidates on election day but have no other connection with the party are members of a coalition only in the loosest possible sense. Their 'cooperation' consists only of voting for certain persons on a certain day; in reality, they are playing more of a noncooperative game, like those who make independent decisions to use metric or imperial measures. But because parties become effective through their parliamentary caucus and party organization, I will regard parties as coalitions for the purposes of this chapter.

Testing the Size Principle

Let us now see whether coalition theory and Riker's size principle provide any new insights into Canadian politics. Table 5.1 contains summary results for the thirteen federal elections held from 1867 through 1917. In these years Canada had a true two-party system. Although there were occasional independent candidates and minor parties, only the Conservatives and Liberals really mattered; between themselves, those two parties typically got at least 97 per cent of the popular vote and all the seats in the House of Commons. The designation 'winning party' in the table means the party that got the most seats – usually, but not always, the party that got the biggest share of the popular vote. The exceptional year was 1896, when the Liberals took 45.1 per cent of the popular vote, 1 per cent less than the Conservatives, but still won more seats than their rivals. 'Margin' refers to the number of percentage points of popular vote by which the winning party led the second party – a negative number in the exceptional case of 1896, positive everywhere else. Finally, note that the Union government, which won the 1917 election, was really the Conservative Party augmented by a large number of breakaway Liberals supporting Robert Borden's war cabinet.

The most striking thing about this array of data is the narrow margin separating the parties in these years. For all thirteen elections, the average margin between the parties was only 4.5 per cent. The only one-sided election came in 1917, when the Liberal split allowed the Unionists to amass a 17.1 per cent margin. In these thirteen elections, the average vote share of the winning party was only 51.1 per cent. Twice the winning party got less than 50 per cent, and five times it ended up with between 50 and 51 per cent – about as close as you can cut it in a two-party system. The effect, however, of Canada's first-past-the-post voting system was to translate these narrow electoral majorities into

TABLE 5.1
Canadian Federal Election Results, 1867–1917

Year	Winning Party	% of Popular Vote	Margin	% of Seats
1867	Conservative	50.1	1.1	60.0
1872	Conservative	49.9	0.8	52.0
1874	Liberal	53.8	8.4	67.0
1878	Conservative	52.5	6.2	68.9
1882	Conservative	50.7	3.9	65.9
1887	Conservative	50.2	1.5	58.6
1891	Conservative	51.1	4.0	56.3
1896	Liberal	45.1	–1.0	55.4
1900	Liberal	51.2	3.8	62.4
1904	Liberal	52.0	5.6	64.5
1908	Liberal	50.4	4.5	61.1
1911	Conservative	50.9	3.2	60.6
1917	Union	57.0	17.1	65.1
Mean =		51.1	4.5	61.4
Standard deviation =		2.5	4.3	4.7

comfortable majorities of seats. The winning party averaged 61.4 per cent of the House of Commons seats in these years.

All of these results are compatible with Riker's view that in politics 'participants create coalitions just as large as they believe will ensure winning and no larger.'[8] Given the stability of the two-party system and the predictable effects of the electoral system, it was rational to build a popular coalition that could win just over half the popular vote. Of course, the politicians of the day did not deliberately calibrate it so finely; they undoubtedly would have been happy to pull more votes and win more seats. Yet regardless of their emotions and intentions, they acted in such a way as to produce a series of minimal winning coalitions.

Table 5.2 extends the time series down to the present, from 1921 through 1997. The reason for presenting a separate table is that throughout this period Canada has had a two-party-plus system. Although the Liberals and Conservatives have continued to form all the governments, other parties – Progressives, CCF/NDP, Social Credit, Reform, and Bloc Québécois – have been able to win significant numbers of seats in every election since 1921. (The Progressive Party existed between 1920 and 1942). The Conservative Party acquired its present name, the Progressive Conservatives, in 1942. As in Table 5.1, 'Winning Party' means the party that took the most seats in the election. (In 1925, the Conservatives got more seats than the Liberals, but not enough for a majority, so the Liberals formed the cabinet with the support of the Progressives.) 'Margin' again means the number of percentage points of popular vote by which the winning party led the second party. There were two cases of negative margin (1957 and 1979), in which the Conservatives got a smaller vote share but more seats than the Liberals.

Several things become apparent in comparing these results to those in Table 5.1. Perhaps most obvious is the greater variability of politics in the second period. In the first period, the winning party's share of

TABLE 5.2
Canadian Federal Election Results, 1921–1993

Year	Winning Party	% of Popular Vote	Margin	% of Seats
1921	Liberal	40.7	10.4	49.4
1925	Conservative	46.5	6.6	47.3
1926	Liberal	46.1	0.8	52.2
1930	Conservative	48.8	3.6	55.9
1935	Liberal	44.8	15.2	70.6
1940	Liberal	51.5	19.8	73.9
1945	Liberal	40.9	13.5	51.0
1949	Liberal	49.5	19.8	73.7
1953	Liberal	48.8	17.8	64.5
1957	Progressive Conservative	38.9	−2.0	42.3
1958	Progressive Conservative	53.6	20.0	78.5
1962	Progressive Conservative	37.3	0.1	43.8
1963	Liberal	41.7	8.9	48.7
1965	Liberal	40.2	7.8	49.4
1968	Liberal	45.5	14.1	58.7
1972	Liberal	38.5	3.5	41.3
1974	Liberal	43.2	7.8	53.4
1979	Progressive Conservative	35.9	−4.2	48.2
1980	Liberal	44.3	11.8	52.1
1984	Progressive Conservative	50.0	22.0	74.8
1988	Progressive Conservative	43.0	11.1	57.3
1993	Liberal	41.3	22.6	60.0
1997	Liberal	38.4	19.0	51.5
Mean =		43.0	10.9	56.5
Standard deviation =		4.8	7.9	10.9

votes ranged from 45.1 to 57.0 per cent, and of *seats* from 52.0 to 68.9 per cent. Compare that to the variation in the second period: from 35.9 to 53.6 per cent for votes, and from 41.3 to 78.5 per cent for seats. Also, the standard deviation of each indicator is appreciably larger in the second period than in the first. This greater range of results is due to the unpredictability of multiparty competition, in which the vote splits between parties can be more important than overall levels of support in determining the share of seats that parties win. Greater uncertainty was also manifested in the prevalence of minority governments in this period. Of the 23 elections held between 1921 and 1997, eight ended without one party winning more than half the seats in the Commons; and only two of the resulting minority governments, those of the Liberals after the 1921 and 1965 elections, lasted more than two years before another election was called.

How did parties react to these new conditions? There were two important changes, both consistent with Riker's size principle. With more than two parties, one need not aim for a majority of the popular vote in order to get a majority of seats; and in the years 1921–97, the winning party averaged only 44.1 per cent of the popular vote and 55.5 per cent of the seats. When the operational definition of minimum winning coalition changed, parties responded by reducing the size of their coalitions. Interestingly, however, the average size of their margin over the second party went up to 10.9 per cent after 1921, as compared to 4.5 per cent before that year. This increase in margin can be seen as a response to the greater uncertainty of multiparty competition. You need a greater margin of error when the erratic effects of vote splits make it harder to calculate results. In multiparty competition, a large lead over the second party can give a comfortable majority of seats even to a party that is far from a majority of the popular vote. This is what happened to the Liberals in 1993, when they turned 41.3 per cent of the popular vote into 60.0 per cent of the seats.

This evidence does not constitute a definitive test of Riker's prediction that 'participants create coalitions just as large as they believe will ensure winning and no larger,'[9] but it is compatible with it. When it became possible to form majority governments with less than 50 per cent of the popular vote, winning parties started to get less than 50 per cent. And when uncertainty increased, winning parties started to increase their margin over their nearest rivals. Other explanations are certainly possible, but nothing so far disqualifies Riker's size principle.

Larger-than-Necessary Coalitions

Riker also argues that larger-than-necessary coalitions are particularly likely to arise in times of great uncertainty, and we have already noted the formation of the Union cabinet during the conscription crisis of World War I. Specifically, Prime Minister Robert Borden proposed the Union government to the Liberals in May 1917, at the same time as he announced his intention to introduce conscription for military service. Sir Wilfrid Laurier, the Liberal leader, knowing that conscription would be extremely unpopular in Quebec, rejected Borden's overture; but it was taken up that fall by a large faction in the Liberal Party, particularly in western Canada. To some extent, Borden forced the Liberals' hand by introducing the *Wartime Elections Act*, which took the vote away from anyone who had been born in one of the enemy countries and naturalized in Canada after 1902. German and Ukrainian immigrants were an important element of Liberal support, especially on the prairies. With the Liberals' prospects diminished by the loss of these voters, many decided to sit with the Conservatives in the Union caucus and cabinet. As soon as this pro-conscription alliance was cemented, Borden struck with an election and received 57 per cent of the popular vote. Laurier's mainline Liberals survived only because of the votes of French-Canadians, who vehemently opposed conscription. The Liberals carried 62 of 65 ridings in Quebec plus the four francophone ridings in New Brunswick, but won only eight of 82 seats in Ontario and two of 57 in the western provinces.[10]

Coalition theory predicts that larger than necessary coalitions should be unstable and prone to break apart. If they were formed in an emergency, the breakup should come quickly once the emergency is over. This was certainly the case with the Union government. The war ended in November 1918; and in June 1919 T.A. Crerar, a leading Liberal Unionist from Manitoba, supported by eight backbenchers, defected from the cabinet when Borden would not abolish protective tariffs on farm machinery. Many Unionist Liberals went over to the new Progressive Party, which favoured free trade, or drifted back to the mainline Liberals, now headed by William Lyon Mackenzie King. The Conservatives were as eager to reclaim full control of the government as the Liberals were to get out of their temporary alliance. The Union officially expired in July 1920, when Arthur Meighen replaced Robert Borden as Conservative leader and prime minister. It had lasted about 20 months.[11]

TABLE 5.3
'Larger-than-Necessary' Winning Coalitions

Year	Winning Party	Popular Vote (Per Cent)	Seats
1958	Progressive Conservative	53.6	78.5
1984	Progressive Conservative	50.0	74.8
1940	Liberal	51.5	73.9
1949	Liberal	49.5	73.7

Apart from the unique case of the Union government, the four largest political coalitions in Canadian history are listed in Table 5.3 in descending order of percentage of seats won; these represent the only four occasions on which the winning party obtained more than 70 per cent of the seats in the House of Commons.

Neither of the two Liberal cases could reasonably be characterized as unstable; both were succeeded by another Liberal majority government in the next election. However, Mackenzie King's coalition of the early 1940s certainly came under great strain. The Co-operative Commonwealth Federation's platform of social justice and the welfare state was so popular that that party actually led the Gallup poll in September 1943, but Mackenzie King headed off the threat by announcing the baby bonus and a policy of full employment.[12] An equally serious threat arose when King resorted to conscription in 1944; if the Conservatives or CCF had had a leader with appeal in Quebec, they might have been able to detach a large share of the French vote from the Liberals. But instead there was a surge of support for independent candidates and small parties (Bloc Populaire, Communists, Social Credit). These scattered forces did take 38.5 per cent of the vote and elect 10 members in 1945, but it was not enough to break the Liberal grip on Quebec. Yet, if circumstances had been different and if King had been a less skilful leader, his large coalition might have broken up rather than merely suffer some decline.

The Diefenbaker sweep of 1958, when the Conservatives won 208 seats, 78.5 per cent of the total, ranks as the greatest triumph in federal political history. He had won a narrow victory in 1957 with 112 seats to the Liberals' 105, formed a minority government, then gone to the polls again within a year. In 1958 his campaign against the Liberals' inexperienced new leader, Lester Pearson, caught fire, and Diefenbaker's popularity soared all across the country, but especially in Quebec, where the Conservatives had won only nine seats with 31.1 per cent of the popular vote the year before.

With the retirement of Louis St Laurent, the Liberals no longer had a French leader; and with the Conservatives in power in Ottawa, arguments based on political opportunism favoured that party and worked against the Liberals in Quebec. Maurice Duplessis, premier of Quebec and leader of the Union Nationale, had always allowed his provincial party to give some assistance to the Conservatives in federal elections; but this time he decided to go all out. It was his chance finally to exact revenge against the federal Liberals for the way they had intervened to defeat the Union Nationale in the provincial election of 1939. Duplessis virtually took over the Conservative campaign in Quebec. He selected 50 winnable ridings out of the 75 in the province and approved $15,000 to be spent in each – a large sum of money at the time. The whole Union Nationale organization was set in motion to help the Conservative campaign. On election day, the Conservatives won every one of the targeted 50 ridings – and only those 50.[13]

Diefenbaker's coalition, however, proved remarkably short-lived – a one-time event rather than the beginning of a new electoral alignment. Duplessis died in 1959, and the Union Nationale was swept from power in 1960 in the election that inaugurated the Quiet Revolution. The Union Nationale machine was broken, and its leaders could no longer draw on provincial government resources to help the federal Conservatives. In the election of 1962, the Conservatives lost 20 percentage points of popular support and fell from 50 seats to only 14 in Quebec. Diefenbaker won enough seats outside Quebec to best the Liberals 118 to 100 and to be able to form a minority government, but it was to be short-lived. The cabinet tore itself apart over the issue of allowing U.S. nuclear warheads on Canadian territory, Diefenbaker was defeated in the House, and the Liberals won a minority government in 1963, commencing a period of 16 years in power. That the greatest electoral coalition in Canadian history disintegrated in only five years is surely consistent with Riker's size principle, even if it is not a definitive test.

Interestingly, the collapse of the Conservatives in Quebec was accompanied by the rise to prominence of Social Credit, which previously had been a fringe player. Under the charismatic leadership of Réal Caouette, Social Credit won 26 seats and 26 per cent of the popular vote. Social Credit was not simply a breakaway fraction of the Conservatives; it drew from both former Liberals and former Conservatives. In one small poll, 17 per cent of those who had voted Liberal and 24 per cent of those who had voted Conservative in 1958 switched to Social Credit in 1962.[14] But losing even a quarter of their support to Social Credit meant that the Conservatives became uncompetitive in many ridings they had won in 1958. The breakup of an electoral coalition does not require massive desertion; it can be accomplished through marginal defections that affect the ability to win seats.

The second-greatest victory in Canadian federal politics was Brian Mulroney's triumph in 1984, when the Conservatives won 211 seats, 74.8 per cent of the total. Again, developments in Quebec were critical to this success. After the Diefenbaker sweep in 1958, Conservative support in Quebec had steadily declined until that party won only one of 75 seats in the 1980 election. Both Robert Stanfield and Joe Clark regarded a breakthrough in Quebec as essential if the Conservatives were ever to return to power, but neither leader was able to achieve it. Clark did form a short-lived minority government after the election of 1979, but he won only four seats in Quebec.

One of Brian Mulroney's main selling points in the contest for the Conservative leadership in 1983 was that he was a native son who could deliver Quebec's vote to the Conservatives.[15] However, he did not at this stage make a particular appeal to Quebec nationalism. In fact, he attacked the 'two nations' and 'community of communities' themes associated with Stanfield and Clark, emphasizing instead a one-Canada approach that hearkened back to Diefenbaker, but also sounded a lot like Pierre Trudeau.[16] Yet once he became leader, Mulroney made an opening to the nationalists. He recruited his law-school friend Lucien Bouchard as a close adviser, and also brought in other well-known supporters of separation, such as Marcel Masse and Benoît Bouchard, who had worked for the Yes side in the 1980 referendum. Going even further, Mulroney fanned into flame the lingering embers of resentment over the patriation of the Constitution in 1982; in a speech written for him by Lucien Bouchard, Mulroney promised to reopen the Constitution, 'to convince the Quebec National Assembly to give its consent to the new Canadian constitution with honour and enthusiasm.'[17]

The strategy worked brilliantly in the short run. In the election of 1984, the Conservatives won a large share of the francophone vote in Quebec and were rewarded with 52 of 75 seats – an astounding increase from their single seat of 1980. Added to the Conservatives' traditional core vote in western Canada and rural Ontario, this was a truly formidable yet highly vulnerable coalition. It was vulnerable because the new nationalist Quebec Conservatives and traditional Conservatives in English Canada had different ideas about such fundamental matters as bilingualism and special status for Quebec. There were two main pillars of the coalition, and it was difficult to keep both happy, because a gain for one was likely to be a loss for the other.

The weakness started to show in 1987, with the foundation of the Reform Party. Significantly, the catalyst for this development was the Mulroney cabinet's decision to grant the billion-dollar maintenance contract for the CF-18 fighter plane to Canadair in Montreal rather than to Bristol Aerospace in Winnipeg, even though the Winnipeg bid was cheaper and judged to be superior on technical grounds. The cabinet's rationale that it was in the 'national interest' to give the contract to a Montreal firm because there was already a concentration of aerospace firms in that city seemed to many in the West merely a hollow rationalization for favouritism to Quebec.[18] Reform drew only 2 per cent of the vote and had little impact upon the 1988 election, but it would be important the next time around.

Meanwhile the clock was ticking on the Meech Lake Accord. This agreement, which was supposed to satisfy Quebec's constitutional demands, had been negotiated in 1987 and had to be approved by all provincial legislatures by 23 June 1990, in order to take effect. Late in 1989, Lucien Bouchard left the cabinet, protesting, he said, the weakness of his colleagues in promoting the Meech Lake Accord. He then proceeded to gather six dissident Conservatives and two Liberals into the Bloc Québécois and turn that organization into a full-fledged party.

The stage was now set for the collapse of the Conservative coalition. It had survived the election of 1988 with a reduced but still comfortable majority, but now it was like a barrel tapped at both ends. In western Canada and rural Ontario, traditional Conservative voters defected to the Reform Party; in Quebec, new Conservative voters, mostly nationalist in orientation, deserted to the BQ. The failures to ratify the Meech Lake and Charlottetown accords accelerated the departures. The day of judgment came on 26 October 1993, when the Conservatives won only two seats in the federal election, against 52 for Reform and 54 for the Bloc Québécois. The Conservatives received 16 per cent of the popular vote, against 19 per cent for Reform and 14 per cent for the Bloc; but the Conservative vote was too thinly spread to be turned into seats. In contrast, the Reform vote was concentrated in the West, especially Alberta and British Columbia, and the BQ vote was concentrated in the

88   Game Theory and Canadian Politics

francophone ridings of Quebec, allowing both parties to achieve local pluralities and elect MPs.

This quick historical review shows some support for Riker's size principle. The Union government lasted less than three years. Of the four largest parliamentary majorities in Canadian history, two fell apart spectacularly within one or two subsequent elections; and one came under severe strain, although it managed to survive. There is no iron law here, but there is clearly some tendency for larger-than-necessary coalitions to disintegrate.

Refinements of the Size Principle

With the introduction of a little more game theory, we can go one step further in specifying which coalitions are most at risk. The classical idea of the solution set and Riker's size principle both use the distribution of benefits to test the rationality of coalitions. Neither pays any attention to the costs of coalition formation; in effect, each assumes that it does not cost anything to put a coalition together. But a moment's reflection shows that that cannot be true in the real world. To build a coalition, one must invest time and other resources in ascertaining the goals of possible partners and in carrying out negotiations to arrive at a common program that holds the coalition together. Thus the desirability of a coalition for rational players must be measured not just in benefits, but in benefits relative to costs.

The political scientist Robert Axelrod, reasoning that the costs of coalition formation will be proportional to the degree of conflict of interest among the partners, set forth two propositions:

1 The less conflict of interest there is in a coalition, the more likely the coalition will form.
2 The less conflict of interest there is in a coalition, the more likely the coalition will have long duration if formed.[19]

For actors, like political parties, that are interested in getting government to do certain things, conflict of interest primarily refers to policy objectives. Thus, writes Axelrod, 'the less dispersion there is in the policy positions of the members of a coalition, the less conflict of interest there is.'[20]

To illustrate the implications of Axelrod's approach, imagine a parliament of 90 seats in which no party has a majority or is even close to it.

FIGURE 5.1
Hypothetical Party Spectrum

Party	(Left)	A	B	C	D	E	(Right)
Number of seats		●	●	●	●	●	
		10	20	20	30	10	

Any government will have to be a coalition government. Imagine further that the parliament contains five parties that can be ordered from left to right on an ideological spectrum. The result is shown in Figure 5.1.

With 90 seats overall, 50 are required to form a majority. If Riker's size principle is interpreted in terms not of the number of parties but of the number of party members involved in the coalitions, the solution set would consist of all possible coalitions of size 50, namely ABC, ADE, BCE, BD, and CD. Riker's theory would have to regard each of these as equally likely on a priori grounds, so it could not make a specific prediction other than the formation of one or another of these coalitions in the solution set.

Against Riker, Axelrod proposes the notion of the minimum *connected* winning coalition, one that 'consists of adjacent members.'[21] The underlying idea is that conflicts of interest are greater among nonadjacent members because their ideologies, and hence their objectives, are further apart. Under these assumptions, the solution set would consist only of ABC and CD because the other minimum winning coalitions would have to reach across internal gaps representing conflicts of interests.

There are many refinements of this approach. For example, some theorists have suggested that the number of parties is also a factor; the fewer the parties, the smaller the bargaining costs. If that is true, we should expect CD rather than ABC. Another refinement is the notion of a compact coalition, which hinges on being able to measure party positions with cardinal numbers, not just ordinal ranks.

Suppose we had a way of measuring party positions that allowed us to draw something like Figure 5.2. Given this set of facts, ABC would seem more likely because it is compact; the possible partners are relatively close to one another. CD would also consist of only 50 members, but C and D are separated by a large gulf that would create many conflicts within the coalition CD.

FIGURE 5.2
Hypothetical Party Spectrum

Party	(Left)	A	B	C		D	E	(Right)
Number of seats		10	20	20		30	10	

It is also possible to conceive of distance between possible coalition partners in more than one dimension, reflecting the reality that issues often cut across one another. Consider the hypothetical situation of Figure 5.3, where the horizontal axis represents a conventional left-right ideological dimension, and the vertical axis represents an ethnic cleavage between, say, the Blues and the Greens.

The parties are arrayed ideologically exactly as they were in Figure 5.2, but now they are also divided on the cross-cutting dimension of ethnic identity. Under these circumstances, the coalition ABC, which is compact ideologically, is now extended on the ethnic dimension, because A is a party of the Greens and B is a party of the Blues. With these positions, any majority coalition will be challenged, because CD is as extended ideologically as ABC is extended ethnically. DE as a minority government, controlling 40 seats out of 90, might be the most stable alignment. In general, any coalition is potentially unstable if the political situation is polarized on more than one dimension, because any

FIGURE 5–3
Hypothetical Party Array

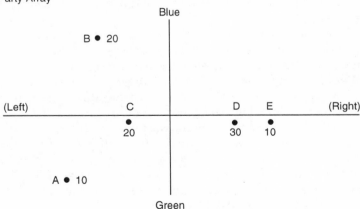

ideological coalition can be broken up by ethnic differences, and any ethnic coalition can be broken up by ideological differences.

Since many European countries are routinely governed by coalition cabinets, there exists a large body of data against which to test different coalition theories. The testing is complicated because there are various ways to formulate both Riker's size principle and Axelrod's distance principle. Without our going into the details, the research seems to show that the size principle in itself is not a good predictor of how cabinet coalitions are formed, but that it works much better when the distance principle is added to it, although it has not been proven that any single definition of distance is the best in all circumstances.[22] One researcher concludes that

parliamentary coalitions can not be explained satisfactorily in terms of the number of seats of the member actors alone: the actors' policy positions must also be taken into account. Theories that take into account the policy positions of the actors without exception, achieve better results than theories that ignore them. And, conversely, theories that achieve a significant outcome all incorporate the policy scale.

The evidence in its entirety strongly suggests that parliamentary majority coalitions in the countries and in the period under study tend to form from actors that are adjacent on the policy scale and, in times of normalcy, those closed coalitions tend to be of minimal range.[23]

The research on cabinet formation is not directly applicable to the topic of this chapter, but it does suggest the importance of policy distance in interpreting political parties as coalitions. It is obvious that both the Diefenbaker and Mulroney coalitions faced large problems of internal distance. In both cases, the Quebec element was brought into the coalition on short notice and was not a traditional part of the Conservative party organization: Diefenbaker's 1958 campaign in Quebec depended on the Union Nationale, and Mulroney brought nationalists and separatists previously associated with the Parti Québécois into the Conservative Party. Also there was in both cases a clash of views. The Union Nationale's long-standing emphasis on autonomy for Quebec was quite different from Diefenbaker's vision of 'One Canada' populated by 'unhyphenated Canadians.' Mulroney, in effect, switched sides to support the nationalist aspirations of members of the Quebec wing of his caucus; but their view, which came to be expressed in the slogan 'distinct society,' was anathema to a great many long-time Conservative voters and party members outside Quebec, especially in western Canada where Diefenbaker's 'One Canada' had been succeeded by the Reform Party's fundamentally similar slogan of 'ten equal provinces.' Both coalitions split apart over this fault line between Quebec and the rest of Canada.

It is instructive that the challenge to Mackenzie King's 1940 Liberal government was of essentially the same type. Conscription of men for military service was far more popular in English Canada than in Quebec; and when King resorted to conscription in 1944, Liberal support dropped sharply in Quebec and a new anti-conscription party, the Bloc Populaire, became a credible contender. But the Liberals were able to hold Quebec. Their share of the popular vote in the province, which had been 63.3 per cent in 1940, fell to only 50.8 per cent in 1945 and bounced back to 60.4 per cent in 1949. Perhaps because of the Liberals' much deeper roots in Quebec, they were able to survive the type of challenge that destroyed Brian Mulroney's Conservative coalition.

In the real world, coalition formation is not a mechanical process. The size principle and the distance principle do not lead automatically to successful predictions about which coalitions will form, survive, or break apart. But size and distance clearly are powerful factors, and defying them is like swimming upstream. You can do it for a while, but it is very tiring and eventually the current carries you away.

6

Who's Got the Power?
Amending the Constitution

In the last week of the 1995 referendum campaign, Prime Minister Jean Chrétien promised to give Quebec a veto over constitutional amendments. Granting this long-standing demand of Quebec politicians was supposed to make Quebeckers more likely to vote against separation in the referendum. After a narrow victory for the No side, the prime minister then had to deliver on his promise. Since a unilateral veto for Quebec would never be acceptable to the rest of Canada, his legal advisers came up with a way of generalizing the veto by granting it to the several major regions of the country. On 27 November 1995, Chrétien released the draft legislation designed to achieve that goal.

Under the draft, the 'general procedure' for constitutional amendment (approval by seven provinces with 50 per cent of the population of the provinces) would have remained in force. However, the federal cabinet would have been barred from introducing a constitutional resolution into the House of Commons unless it already had the approval of at least Ontario, Quebec, two provinces in the West having more than 50 per cent of the population of that region, and two Atlantic provinces having more than 50 per cent of the population of that region.[1] This arrangement was popularly described as 'lending' the federal government's veto to Quebec (and the other designated regions). On 7 December, responding to complaints from British Columbia, the minister of justice announced that this four-region veto would be changed to a five-region formula under which passage of an amendment would require approval by Ontario, Quebec, British Columbia, two of three prairie provinces having more than 50 per cent of the population of the prairies, and two of four Atlantic provinces having more than 50 per cent of the population of that region.

Using party discipline and time allocation, the Liberal government passed its legislation quickly through the House of Commons, and the Senate did not stand in the way. However, there was heavy criticism of the bill, particularly in British Columbia and Alberta, and also among native leaders. Nor did the legislation appear to satisfy Quebec's provincial politicians, who denounced it as too little, too late. The federal minister of justice talked about further entrenching the five-region veto in the Constitution, but that prospect seems unlikely, at least for the time being. When the first ministers met on 21 June 1996, they avoided all talk of constitutional amendments. The five-region veto would be particularly difficult to constitutionalize because, as an amendment to the amending formula, it would require the unanimous approval of all provinces in addition to Parliament.[2]

This chapter attempts to shed light on the controversy by using the Banzhaf Power Index (BPI) to analyse the existing and proposed amending formulas.[3] The first conclusion is that the regional veto will create tiers of provinces with very different and permanently unequal shares of amending power. Thus, if we regard even rough equality of the provinces as a value to be protected, there is much to be said for the existing 7/50 formula. The second conclusion is that the regional veto, as it redistributes amending power from some provinces to others, also makes all provinces worse off in terms of their ability to obtain future changes to the Constitution.

The Banzhaf Power Index (BPI)

In the preceding chapter, we looked at coalitions in which all members have an equal share of decision-making power. But both the general procedure for amending the Canadian Constitution and the new five-region veto are n-person voting games in which the participants – the provinces – have differing shares of power, depending upon the size of their population. We have to introduce another methodology, known as the Banzhaf Power Index, for analysing such weighted voting games.

Weighted voting systems are not common in Canadian politics. Other than the constitutional amending procedures discussed here, the main example would be the Canada Pension Plan, changes to which require the approval of two-thirds of the provinces holding two-thirds of the population.[4] However, they are sometimes found in the realm of international organization, where states of widely differing size and resources cooperate in decision-making. Relevant examples would be the 15-member United Nations Security Council, where only the five permanent members have a veto, and the European Union's Council of Ministers, where voting strength is roughly proportional to population. Outside the world of government, weighted voting is commonplace in the annual meetings of corporations, where owners have widely different shareholdings and the voting rule is not 'one owner, one vote,' but 'one share, one vote.' This difference in voting strength may carry over to the board of directors, where varying sized blocs of directors often represent the interests of the major owners.

The BPI is a method for comparing the relative decision-making power of participants in systems of weighted voting. Steven J. Brams defines the BPI as 'the number of winning coalitions in which the member's defection from the coalition would render it losing – which is a critical defection – divided by the total number of critical defections for all members.'[5] Winning coalitions that would be turned into losing coalitions by the defection of a single member are known as minimum winning coalitions, or MWCs; we have met them before. Thus, the BPI is computed by enumerating all MWCs, counting the number of critical defections (*pivots*) for each player, and dividing by the sum of pivots for all players.

As an example, consider a hypothetical system with four participants – A, B, C, and D. A has three votes, B has two, and C and D each have one. There are seven votes in total, so any MWC must have at least four votes if the decision rule is simple majority. All MWCs are tabulated below in the upper panel of Table 6.1; members that are pivotal in a particular coalition are printed in boldface.

It might seem obvious that A, with three votes, would have the greatest voting power; but it is not obvious that the power of B, who has two votes, should be equal to that of C and D, who have only one vote each. This surprising result is a function of the coalitional possibilities in the particular situation; it would not necessarily be true in a different configuration of voters and weights. The virtue of the BPI is that it focuses precisely on particular configurations and thus moves beyond generalities based on surface impressions.

The BPI can also be used to measure voting power where the decision rule is one of qualified majority, as in the 7/50 amending formula. For a simple illustration, see Table 6.2. In it, everything remains the same as in the previous example, except for the decision rule, which is now set at five votes out of seven.

TABLE 6.1
Sample Calculation of BPI in a Weighted-Voting Game

Enumeration of MWCs				
n = 2	**AB**	**AC**	**AD**	
n = 3	ABC	ABD	**BCD**	ACD
n = 4	none			

Enumeration of Pivots		
Player	Number of Pivots	BPI
A	6	.50
B	2	.17
C	2	.17
D	2	.17
Total	12	1.00

As it happens, the winner in this move from simple to qualified majority is B, whose BPI increases from 0.17 to 0.30, while C and D fall from 0.17 to 0.10 and A remains unchanged at 0.50. This shift illus-

TABLE 6.2
Calculation of BPI in a Weighted-Voting Game with Qualified-Majority Rule

Enumeration of MWCs			
n = 2	**AB**		
n = 3	**ABC**	**ABD**	**ACD**
n = 4	**ABCD**		

Enumeration of Pivots		
Player	Number of Pivots	BPI
A	5	.50
B	3	.30
C	1	.10
D	1	.10
Total	10	1.00

trates a fundamental principle of BPI analysis that will become strikingly evident in the discussion of constitutional amending formulas: because power is a relational concept, any change in the number of players, weighting of votes, or decision rules can have unanticipated and perhaps undesired consequences in the distribution of power in the system.

Finally, note that for any configuration there are not only minimum winning coalitions but also minimum blocking coalitions (MBCs). That is, if it takes five out seven votes to pass a measure under a qualified majority rule, three votes can block it. The BPI will be the same whether the analysis is done in terms of MWCs or MBCs, as illustrated in Table 6.3. Depending on the situation, it is sometimes more practical to compute the BPIs in terms of MBCs rather than MWCs; but it is always possible to do it both ways and get the same results.

The 7/50 Formula

The then new 7/50 rule was analysed by the mathematician D. Marc Kilgour in 1983, using population statistics from the 1981 census. Kilgour proceeded by way of MBCs. In his words:

TABLE 6.3
Calculation of BPI in Terms of Minimum Blocking Coalitions

Enumeration of MBCs

n = 1	A				
n = 2	AB	AC	AD	BC	BD
n = 3	ACD	BCD			
n = 4	none				

Enumeration of Pivots

Player	Number of Pivots	BPI
A	5	.50
B	3	.30
C	1	.10
D	1	.10
Total	10	1.00

98   Game Theory and Canadian Politics

(1) No province acting alone can block an amendment.
(2) Of the 45 possible coalitions of two provinces, only the coalition of Ontario and Quebec can block an amendment.
(3) Of the 120 possible coalitions of three provinces, only 12 have the power to block an amendment. These are the eight which include both Ontario and Quebec, and the four consisting of Ontario and British Columbia together with one of Alberta, Saskatchewan, Manitoba, Nova Scotia.
(4) Every coalition of four or more provinces can block an amendment.[6]

Using this approach, Kilgour calculated the following BPIs for the provinces, as shown in Table 6.4.[7]

According to Kilgour's calculations, the 7/50 rule would treat the provinces in relatively equal fashion, giving Ontario, the strongest province, only 33 per cent more power than each of the three weakest provinces (0.1234/0.0929 = 1.33). Moreover, it did not classify provinces into permanent categories differentiated by possession or non-possession of a veto. No province had a veto by itself, and all provinces could help exercise a veto by becoming pivotal partners in a variety of MBCs.

Note that the provinces' BPIs change over time as population changes. These are gradual adjustments registered in the census every 10 years, and usually they affect only the third or fourth decimal place of the BPI. Occasionally, however, they may have a greater impact. For example,

TABLE 6.4
Kilgour's BPIs for the 7/50 Rule, 1981 Census

Province	Population (millions)	BPI
Ontario	8.60	.1234
Quebec	6.41	.1132
British Columbia	2.72	.1031
Alberta	2.20	.0954
Saskatchewan	.96	.0954
Manitoba	1.02	.0954
Nova Scotia	.85	.0954
New Brunswick	.70	.0929
Newfoundland	.57	.0929
Prince Edward Island	.12	.0929
Total	24.22	1.0000

by 1995 British Columbia's population had grown to the point that its BPI, if calculated on the basis of that year's estimated population statistics, would have equalled that of Quebec. The change would not take legal effect until the next general census,[8] but the writing was on the wall. Indeed, it might seem unfair that Quebec, whose population was still more than twice as great as that of British Columbia, would have only the same share of power as the latter province. However, it is not humanly possible to contrive a formula that will be exactly fair to 10 provinces with different and ever-changing populations. No doubt each province could find that it is treated unfairly relative to one or more other provinces. Nonetheless, the 7/50 rule stands out as an ingenious compromise between provincial equality, demographic weight, and feasibility in operation.

The Five-Region Veto

The five-region veto found in the final version of the Liberals' legislation always requires a MWC of size 7 or greater. A successful amendment must have the support of at least Ontario, Quebec, and British Columbia, plus Alberta with either Saskatchewan or Manitoba, plus two out of three of Nova Scotia, New Brunswick, and Newfoundland. Three initial observations can be made.

First, the five-region veto completely supersedes the 7/50 rule inasmuch as any coalition of seven provinces that meets the five-region test will also have more than 50 per cent of the population of Canada. Thus, the five-region formula can be analysed on its own without our worrying about interaction with the still constitutionally valid general procedure.

The second observation is that Prince Edward Island becomes a dummy player, deprived of all power under this formula. Its population is so small that it can never be pivotal in the Atlantic region; and since any MWC under this formula already contains seven provinces, PEI can never be pivotal in that sense. The disenfranchisement of PEI was probably not intended, but it is nonetheless complete.

In a well-intentioned but misguided attempt to rescue PEI, Nova Scotia and New Brunswick announced they would lend their vetoes to the island province. Each said that, if PEI were to join with the other, it would also join the coalition, making it a MWC for the region.[9] For example, if New Brunswick and PEI were initially opposed to an amendment, Nova Scotia would be obliged also to oppose it, thus causing its defeat by a coalition of three Atlantic provinces with more than 50 per cent of the regional population. This unenforceable promise would indeed rescue PEI, but it would effectively disenfranchise Newfoundland; for there would now be no MWC in which Newfoundland could be pivotal. If Newfoundland joined with New Brunswick and Nova Scotia, the coalition would be larger than minimal, and none of the three would be pivotal. If Newfoundland joined PEI, the coalition would be a losing one because New Brunswick and Nova Scotia together have more than half the population. And Newfoundland could not join Nova Scotia or New Brunswick alone, because each would be pledged to join the other along with PEI.

The third observation is that in the 1996 census, Alberta had 2.70 million people, against 1.11 million for Manitoba and 0.99 million for Saskatchewan. Alberta thus has a de facto veto because it has more than half the population of the three prairie provinces; and because its population is growing, that veto is likely to persist for the foreseeable future. On the other hand, Saskatchewan and Manitoba do not become powerless dummies like PEI. Alberta can veto an amendment by itself, but it cannot help pass an amendment without the support of either Manitoba or Saskatchewan. The two smaller prairie provinces share a veto, so to speak.

The mathematical analysis is straightforward. There are only six possible MWCs of size 7. Ontario, Quebec, British Columbia, and Alberta are always pivotal; Saskatchewan and Manitoba are each pivotal three times; and Nova Scotia, New Brunswick, and Newfoundland are each pivotal four times. For $n = 8$, there are 11 MWCs. If Saskatchewan and Manitoba are both added to the essential four, the remaining two from the Atlantic region must be two out of the three of Nova Scotia, New Brunswick, and Newfoundland. If Manitoba or Saskatchewan is missing, there must be some combination of three out of four Atlantic provinces. There are only six MWCs of size 9 because the four veto-wielding provinces must always be included, and of course there is only one MWC of size 10.

Summing the pivots for MWCs of all sizes yields the results shown in Table 6.5.

As mentioned above, Prince Edward Island is the big loser under this scheme, becoming a powerless dummy. The other Atlantic provinces and the two smaller prairie provinces also incur significant losses compared to their status in the 7/50 rule, while the four veto-wielding provinces make big gains. Ironically for a measure that was sup-

TABLE 6.5
BPIs for the Five-Region Veto, 1996 Census

Province	Population (millions)	BPI	Ratio over 7/50 BPI
Ontario	10.75	.1622	131
Quebec	7.14	.1622	143
British Columbia	3.72	.1622	157
Alberta	2.70	.1622	170
Saskatchewan	.99	.0541	57
Manitoba	1.11	.0541	57
Nova Scotia	.91	.0811	85
New Brunswick	.74	.0811	87
Newfoundland	.55	.0811	87
Prince Edward Island	.13	.0000	0
Totals	28.74	1.000	

posed to respond to demands from Quebec, the biggest winner is Alberta (70 per cent increase) and the second-biggest is British Columbia (57 per cent increase).

Everyone Loses

Even though the four big provinces are all 'winners' under the five-region formula, they are winners primarily in the negative sense of being able to prevent changes desired by others. Their increased BPI does not imply an increased ability to secure changes that they themselves might wish. BPI analysis shows that the four big provinces are now more powerful relative to the six small provinces in the constitutional amendment process, but all provinces are less powerful in an absolute sense of being able to change Canada constructively in the future. The regional-veto formula decreases the horizon of possibilities for all players, even as it reallocates power from player to player.

We can measure the contraction of the horizon by enumerating the number of winning coalitions (all winning coalitions, not just MWCs) for any amendment formula and dividing by the total number of possible coalitions. With 10 provinces able to vote yes or no, the total number of possible coalitions is, according to the binomial expansion, $2^{10} = 1024$, ranging from 10 yes and 0 no to 10 no and 0 yes. The least restrictive formula that anyone would recommend for amending a con-

TABLE 6.6
Winning Coalitions for Different Amending Formulas

Definition	Number	% of Total
All	1024	100.0
Simple majority	386	37.7
7/50 rule, 1981 census	163	15.9
7/50 rule, 1991 census	161	15.7
5-region veto	31	3.0
Unanimity	1	0.1

stitution would be a simple majority, in Canada's case six of 10 provinces. Under that rule, 386 coalitions, 37.7 per cent of the total, would become winning coalitions – 210 of size 6, 120 of size 7, 45 of size 8, 10 of size 9, and one of size 10. Under a similar approach, the number of winning coalitions for other rules is given in Table 6.6.

The general procedure, allowing positive results in about 16 per cent of cases, is certainly more restrictive than a simple majority; indeed, as a qualified-majority rule, it is designed to be so. However, it still holds out a realistic possibility of achieving amendments. The five-region veto, on the other hand, is much closer in practice to unanimity than it is to the general procedure; it allows only 31 provincial coalitions – 3 per cent of the total – to become winning coalitions.

A measure of each province's positive power to achieve a desired constitutional amendment can be derived by multiplying its relative share of power (BPI) under a particular formula times the probability for all participants together to achieve an amendment. Table 6.7 compares each province's positive power in 1981 under the 7/50 formula against its power under the five-region veto, using 1996 census figures. The table shows that even provinces like Alberta, which have more *relative* power under the five-region veto, now have less *absolute* power to achieve desired constitutional amendments.

The effect of the five-region veto is to freeze the constitutional order around the status quo. This result is bound to be unpopular in the West, where ideas such as Senate reform, property rights, and a Canadian common market still have many adherents. The five-region veto may be more popular in Quebec if it assuages fears of a repetition of 1981–2, in which the Constitution was amended without Quebec's approval. However, politicians in Quebec, federalists as well as separatists, have positive constitutional aspirations that go beyond forestalling

TABLE 6.7
Positive Power Indexes

Province	7/50 Procedure (1981)	Five-Region Veto (1996)
Ontario	.0196	.0049
Quebec	.0180	.0049
British Columbia	.0164	.0049
Alberta	.0152	.0049
Saskatchewan	.0152	.0016
Manitoba	.0152	.0016
Nova Scotia	.0152	.0024
New Brunswick	.0148	.0024
Newfoundland	.0148	.0024
Prince Edward Island	.0148	.0000

changes desired by others. If done under the rule of law, a transfer of jurisdictions from Ottawa to Quebec City, as well as the separation of Quebec from Canada, will involve constitutional amendments that will now be harder to obtain than they were previously. Ironically, therefore, it may be Quebec, as the province least satisfied with the constitutional status quo, that loses most from the five-region veto.

In a sense, this prediction had come true by the middle of 1996. Early that year, Prime Minister Jean Chrétien had his Liberal majority in the House of Commons pass a resolution pledging the House to treat Quebec as a distinct society; and he wanted to go further and put a distinct society clause in the Constitution.[10] But by June 1966 all movement toward amending the Constitution had been abandoned, at least for the time being, when it became clear to the prime minister that provincial support was simply not forthcoming. Under the 7/50 rule, it might have been conceivable to get agreement from Ontario, Saskatchewan or Manitoba, and the four Atlantic provinces, and then go to Quebec, challenging that province to become the seventh supporter. At least then the blame would clearly fall upon the separatist government of Quebec for causing the distinct society amendment to fail. But with the five-region veto in place, both British Columbia and Alberta, where the idea of distinct society is highly unpopular, now had the power to veto the prime minister's project.

The prime minister and his legal advisers perhaps did not think through all the legal implications of beginning with the five-region veto. If they were really intent upon getting distinct society, they should have held off on the five-region veto until after making a serious effort to get distinct society under the general procedure.[11] The Canadian constitutional game has now become so technically complex that otherwise rational actors may make outright mistakes about how to pursue their own objectives.

On the other hand, there is still one avenue open for those seeking to entrench distinct society in the Constitution. The five-region veto prohibits the *government* – that is, the cabinet – from introducing a constitutional resolution into Parliament without first getting the support of the five regions, but it does not prohibit a private member from introducing such a resolution. Daniel Johnson, the former leader of the Quebec provincial Liberals, proposed getting six provinces outside Quebec to support distinct society and then challenging Quebec to become the seventh. This approach would constitute a political strategy for fighting a future Quebec provincial election as well as a way to amend the Constitution. In this scenario, approval from Parliament would presumably come through passage of a private member's resolution. Technically, this plan could work, but there would almost certainly be a swell of outrage in Alberta and British Columbia if the Liberal government allowed such a transparent evasion of its own legislation. Fearing such a backlash, the government may be reluctant to use the technical escape allowed by the law. Or maybe the federal government will decide, as it has occasionally in the past, that opinion in the west doesn't matter in comparison to the momentous concerns of Quebec.

7

The 'Right Stuff': Choosing a Party Leader

Condorcet Winners and Cycles

The leader of a political party needs to have wide support throughout the organization. To win an election, donors have to give money, activists have to work in the campaign, and voters have to turn out on election day. Any or all of those efforts could be hampered if large segments of the party dislike the leader. This situation is obviously true in a U.S.-style presidential system, where popular votes are cast directly for the party's leader as candidate for president; and it is equally true in a modern parliamentary system like Canada's, where the pervasive influence of the mass media means that voters identify the leader with the party and are much less swayed by the merits of local candidates.

The system of first-past-the-post or plurality voting used in Canadian elections would not function well for choosing a party leader. If there were more than two candidates and the victory simply went to whoever got the most votes on a single ballot, a leader might be selected with the support of only 40 or 30 per cent, or perhaps even fewer, of those doing the choosing. In this situation, a rational-choice theorist would prescribe selection of a *Condorcet winner* as the best method of ensuring wide support.

The term Condorcet winner is named after the Marquis de Condorcet, an 18th-century mathematician and philosopher who was an advocate of the French Revolution but got on the wrong side politically and died in prison after being arrested. His work on voting procedures did not have any immediate influence, but he is now recognized as a pioneer of rational-choice theory. In the Condorcet procedure, the candidates run against each other in all possible pairwise combinations, and the one who beats all the others is declared the winner – rather like in a round-robin curling bonspiel or hockey tournament. There is a compelling case for regarding such a Condorcet winner as the best choice, for by definition a Condorcet winner is preferred by the voters to all other alternatives.[1]

Someone who could get a simple majority of 50 per cent + 1 of the votes on the first ballot would automatically be a Condorcet winner, but not all Condorcet winners would be majority winners.

Consider the profile of support shown in Table 7.1. No one commands a majority of first preferences, but b is the Condorcet winner; that is shown in Table 7.2, which tabulates the vote for all pairwise contests. Read Table 7.2 horizontally. In the row labelled a, the entry in the column labelled b represents the percentage of the vote that a would get against b, and so on. The Condorcet winner is b because b gets 64 per cent of the vote against a and 66 per cent against c. A plurality contest would be won by a, and yet a is less widely supported because he is the last choice of 34 per cent, whereas b is not the last choice of anyone. In common-sense terms, b would make the best party leader because he is the first choice of 30 per cent and the second choice of the other 70 per cent of voters. Yet he would come last in a first-past-the-post contest because a is the first choice of 36 per cent and c is the first choice of 34 per cent of voters.

TABLE 7.1
Hypothetical Preference Orderings of Three Groups over Three Options

Group 1 (36% of voters)	a > b > c
Group 2 (34% of voters)	c > b > a
Group 3 (30% of voters)	b > a > c

TABLE 7.2
Pairwise Comparisons Based on Preferences of Table 7.1

	a	b (per cent)	c
a	–	36	66
b	64	–	66
c	34	34	–

TABLE 7.3
Example of a Cyclical Preference Structure

Preference Orderings		
Group 1 (36%)	a > b > c	
Group 2 (34%)	c > a > b	
Group 3 (30%)	b > c > a	

Pairwise Comparisons			
	a	b (per cent)	c
a	–	70	36
b	30	–	66
c	64	34	–

The Condorcet procedure is a beautiful theoretical construct, but it has two grave flaws as a practical device. First, it quickly becomes cumbersome as the number of alternatives increases. If there were six candidates for leader – not unusual in the real world – those responsible for making the choice would have to vote on (6 x 5)/2 = 15 different pairwise comparisons. Second, and even more objectionable, the existence of a Condorcet winner cannot be guaranteed. In the situation known as a *cycle*, there is no alternative that is preferred to all the others.

Consider now the preference profiles shown and analysed in Table 7.3. In this profile, *a* defeats *b*, and *b* defeats *c*, but *c* defeats *a*. That is, every alternative can be defeated by another alternative. If we were to write a collective preference ordering, we would have to put down *a* > *b* > *c* > *a*, which illustrates why such an example is called a cycle: the collective preferences take you in a circle back to the starting point.

A cyclical preference ordering seems to violate the rational-choice axiom of transitivity; it says simultaneously that *a* is preferred to *c* and that *c* is preferred to *a*. In fact, the axiom of transitivity applies only to individual preferences. Collective preference orderings generated by aggregation of individual preferences often violate transitivity, so groups are not rational in the same sense as individuals. Another term for this phenomenon is the 'paradox of voting,' which captures the fact that the aggregation of individually rational and transitive preference orderings can produce a collectively intransitive, and hence irrational, preference ordering.

Nominating Conventions

If the group of those choosing a party leader had a cyclical preference structure as shown in Table 7.3, they could never come to a conclusion if they were using the Condorcet procedure – and yet a party must have a leader. A collective choice rule that cannot guarantee an outcome is hardly practical for a leadership contest. What is needed is a practical rule that promotes consensus on the choice of leader. In the United States, the Democratic Party, which began in 1832 to choose its presidential candidate at a national nominating convention, adhered until 1936 to a successive-ballot procedure requiring a two-thirds majority. This practice guaranteed that no candidate could be selected without wide support, but the two-thirds threshold was sometimes almost impossible to meet. In 1924, for example, the delegates voted an incredible 103 times before reaching a conclusion.[2] Since 1936, the Democrats have adopted the procedure that the Republicans had adhered to ever since their first convention in 1856, namely successive ballots with a simple majority (50 per cent +1) decision requirement.

The first national Canadian nominating convention, held by the Liberals in 1919, adopted the Republican procedure, but with a slight modification. The change made by the Liberals was that, if no one had received a majority after four ballots, the trailing candidate would be required to withdraw on the fifth and each subsequent ballot until one received a majority.[3] This change from a successive ballot to a true runoff guarantees a reasonably speedy ending. If there are n candidates, there can be no more than two candidates left at the $(n-1)$ ballot; and in a choice between two options, one must necessarily receive a majority. Since that beginning, all Canadian parties have followed the same general principle but have gradually tightened up the rules about dropping trailing candidates. Each party has its own rules, which vary slightly; but in general the dropping now begins after the first ballot and includes not just the candidate with the lowest total but any others who fall below some very low threshold, such as 75 votes. Such rules virtually guarantee that a decision will be reached after three or four ballots at most, and the delegates will be able to go home on schedule. In our age of inflexible hotel and airplane reservations, it would scarcely be practical to hold two or three thousand people indefinitely while they cast ballot after ballot.

The now standard Canadian procedure for selecting party leaders at a national convention scores well on efficiency and practicality, but there is still a nagging question about its effectiveness. That is, does it succeed in choosing the candidate with the widest support in the party? More precisely, does it sometimes select someone other than the Condorcet winner, even if there is a Condorcet winner?

It is not difficult to show that this is a theoretical possibility. In fact, using the Canadian run-off procedure with the profile of support shown in Table 7.1 would result in a victory for a even though b is the Condorcet winner. Candidate b would be eliminated after the first ballot because he would be last of the three candidates in terms of first preferences. If b's supporters act rationally on the second ballot by voting for their second preference, they will vote for a, and a will defeat c, 66 against 34 per cent. The general problem is that the run-off procedure can eliminate a potential Condorcet winner before the final vote is taken.[4]

How probable is this theoretical possibility? Certainly, there are many situations in which it is not worth worrying about. If there is one strong candidate who easily gets a majority on the first ballot, as did Louis St Laurent at the Liberal convention in 1948 and John Diefenbaker at the Conservative convention of 1956, there is no problem, because a first-ballot majority winner must also be a Condorcet winner. The problem of missing a Condorcet winner is most likely to arise under certain special conditions resembling the profile of Table 7.1:

- there are three or more serious candidates;
- no one is close to a first-ballot majority;
- the support of the two front runners is polarized – that is, those who have a first preference for the one tend to have a last, or at least a very low, preference for the other; and
- a candidate who is third or even lower in terms of first preferences might be a Condorcet winner because she is the second choice of many of those who support one of the two front runners.

Under such conditions, there is a definite risk that the run-off procedure might eliminate a potential Condorcet winner who would be best able to unify the party. Although these conditions are not typical of Canadian leadership contests, neither are they unknown. Let us, therefore, look more closely at three recent leadership races that bear some resemblance to this model.

The Progressive Conservatives, 1983

On 31 January 1983, at a national Conservative meeting in Winnipeg, 33.1 per cent of the delegates voted to hold a convention to choose a

110 Game Theory and Canadian Politics

TABLE 7.4
Voting at the 1983 Conservative Leadership Convention

Candidate	First ballot	Second ballot	Third ballot	Fourth ballot
Brian Mulroney	874	1021	1036	1584
Joe Clark	1091	1085	1058	1325
John Crosbie	639	781	858	–
David Crombie	116	–	–	–
Michael Wilson	144	–	–	–
Peter Pocklington	102	–	–	–
John Gamble	17	–	–	–
John Fraser	5	–	–	–
Total	2988	2887	2952	2909

new leader. Thereupon, Joe Clark, who had been leader of the party since 1976 and had been undermined by internal opposition for much of that time, announced that he would resign and enter a new race, seeking to be re-elected. His main opponent was Brian Mulroney, who had been working behind the scenes for years to bring Clark down; but the Newfoundland MP John Crosbie also emerged as a serious contender.[5]

When the leadership convention took place in Ottawa in June 1983, there were eight candidates, and it took four ballots to choose Brian Mulroney as the new leader. The results of the four ballots are shown in Table 7.4.[6]

This was clearly a three-man race. Crosbie, the weakest of the top three, was more than 500 votes ahead of the other five on the first ballot; and those bottom five candidates had fewer than 400 votes among themselves. Among the top three, Clark started the strongest but lost a little on the second and third ballots. Mulroney and Crosbie both increased on the second and third ballots; but under the iron Darwinian logic of the Canadian run-off system, Crosbie had to drop out after the third ballot, setting the stage for Mulroney's victory.

Soon afterward, political analyst Terence Levesque raised the question of whether the delegates had selected the 'right' leader – 'right' in the sense of being a Condorcet winner. When the delegates had a chance to make a pairwise comparison of Mulroney and Clark, they chose Mulroney; so Clark could not have been a Condorcet winner. However, the run-off gave them no chance to compare Crosbie one-on-one with either Mulroney or Clark, leaving open the possibility that Crosbie might have been a Condorcet winner. Levesque argued that Crosbie was in fact the 'right' choice in this sense.

The situation meets the conditions specified above. There were three major candidates, no one was close to winning on the first ballot, and the support of the top two contenders was strongly polarized. Mulroney had been in the race in 1976 when Clark had won; he had intrigued for years to weaken Clark's position as leader; and the two sides had waged a dramatic struggle in this campaign, particularly in Quebec, to get delegates on their side. It was not unreasonable to think that many Clark and Mulroney supporters might prefer Crosbie over their main antagonist; and if enough felt that way, Crosbie could be a Condorcet winner.

An argument based on electoral arithmetic can be made to support this conclusion. There were 2952 votes cast on the third ballot, so a majority of 50 per cent + 1 was 1477. Crosbie received 858 votes in the three-way competition on this ballot, so he would have needed an additional $1477 - 858 = 619$ votes to win a head-to-head contest against Clark or Mulroney. Could Crosbie count on recruiting 619 votes from the Clark camp in a contest against Mulroney? Clark received 1058 votes on the third ballot; the 619 that Crosbie needed would constitute $619/1058 = 58.6$ per cent of Clark's support. Similarly, to win a head-to-head contest against Clark, Crosbie would have needed to recruit $619/1036 = 59.7$ per cent of Mulroney's support.[7] It seems plausible that, in view of the well-known polarization between the Clark and Mulroney forces, 60 per cent of each side would have chosen Crosbie over their main antagonist; and if that was true in both cases, Crosbie would have been the 'right' choice – that is, a Condorcet winner prematurely eliminated by the mechanics of the run-off system.

'Plausible,' however, is not the same as certain. In response to Levesque, George Perlin drew on three surveys he had administered to convention delegates. In the third and final survey, administered after the convention was over, he asked the respondents ($n = 755$) how they would have voted in two-way contests between Crosbie and Clark, and Crosbie and Mulroney. Crosbie did indeed defeat Clark 52 to 45 per cent, with 3 per cent abstaining; but he lost to Mulroney 40 to 54 per cent, with 6 per cent abstaining.[8] If these data can be generalized to the whole group of voting delegates, Mulroney was the Condorcet winner, able to defeat both Clark and Crosbie in direct pairings, so the run-off system worked in this case to select the 'right' candidate.

112 Game Theory and Canadian Politics

It is true, Perlin points out, that many of Clark's followers had great antipathy toward Mulroney; but they were even more reluctant to vote for the unilingual Crosbie. Joe Clark had made reconciliation with Quebec a hallmark of his political career; and many of his followers, even though they resented Mulroney's disloyalty, saw him as more able than Crosbie to carry on in the same vein.[9]

The New Democrats, 1989

After Ed Broadbent announced his retirement as NDP leader, there was a vigorous contest to succeed him, resulting in the selection of Audrey McLaughlin as the first female leader of a major federal party in Canada. Table 7.5 summarizes the voting on the ballots at the December 1989 leadership convention.[10]

These results do not give a surface impression that the run-off procedure failed to select the 'right' leader. From the beginning, there were not three but only two front runners, McLaughlin and Barrett. Steven Langdon, who finished third on the first ballot with 351 votes, was more than 200 votes behind the second-place Barrett and only 36 votes ahead of the fourth-place Simon de Jong. Langdon did move ahead on the second ballot, but then fell back sharply on the third. Although it is a mathematical possibility, it is hard to believe he was more widely acceptable to the delegates than both McLaughlin and Barrett.

That McLaughlin was the Condorcet winner can be demonstrated from a survey of convention delegates conducted by Keith Archer, who

TABLE 7.5
Voting at the 1989 NDP Leadership Convention

Candidate	First ballot	Second ballot	Third ballot	Fourth ballot
Audrey McLaughlin	646	829	1072	1316
Dave Barrett	566	780	947	1072
Steven Langdon	351	519	393	-
Simon de Jong	315	289	-	-
Howard McCurdy	256	-	-	-
Ian Waddell	213	-	-	-
Roger Lagassé	53	-	-	-
Total	2400	2417	2412	2388

TABLE 7.6
Pairwise Comparisons of Leading Candidates in 1989 NDP Leadership Contest (per cent)

	McLaughlin	Barrett	Langdon
McLaughlin	–	62	66
Barrett	38	–	43
Langdon	34	57	–

asked respondents to rank all the candidates from first down to seventh choice.[11] Such rankings can be used to simulate a Condorcet round-robin contest. According to Archer's data, each of the top three – McLaughlin, Barrett, and Langdon – would have defeated all of the other four candidates. Moreover, McLaughlin would have defeated Barrett and Langdon, as shown in Table 7.6. It could be considered a defect of the run-off system that, although Langdon would have defeated Barrett in the Condorcet procedure, it was Langdon, not Barrett, who was forced out of the race after the third ballot. But this minor defect did not affect the final result; McLaughlin, the Condorcet winner, also won the race at the convention.

The New Democrats, 1995

After the New Democrats' electoral disaster of 1993, in which, with only 6.9 per cent of the popular vote and nine seats, the party did not even win official-party status in the House of Commons, Audrey McLaughlin resigned as leader. In an attempt to rebuild the party's grassroots organization, the NDP decided to experiment with the leadership selection process. In front of the national convention required by the party's constitution, it tacked on a sort of primary system. There would be five ballots open to party members in the five regions of the country – British Columbia, the Prairies, Ontario, Quebec, and Atlantic Canada – plus a sixth primary for affiliated union members. To get on the ballot for the October convention, a candidate would have to win at least one primary race or get more than 15 per cent of the votes cast in all the primaries together. Although the primaries would give some indication of candidate support, the delegates would not be bound by the results and could vote as they wished when they got to the convention floor in Ottawa.[12]

Three candidates made it to Ottawa. British Columbia MP Svend Robinson won three primaries – British Columbia, Ontario, and Quebec – and 32 per cent of all votes cast in the primaries. Lorne Nystrom, a long-serving MP from Saskatchewan who had lost his seat in 1993, won two primaries – the Prairies and labour – and 47 per cent of all votes cast. Alexa McDonough, leader of the NDP in Nova Scotia and member of that province's legislature – won the Atlantic primary and 19 per cent of votes overall. Because of the primary results, many observers ranked Nystrom first, Robinson second, and McDonough third; but they would have done well to pay attention to Manitoba NDP leader Gary Doer's prescient remark that 'Alexa is everybody's second choice.'[13]

The three candidates had quite different bases of support. As an out-of-the-closet gay, an environmentalist who had been arrested for his activism, and an advocate of euthanasia, Robinson had a strong following among 'single issue' feminists, gays, and environmentalists. Many of his young supporters were working in their first leadership race. Nystrom, associated with Saskatchewan's Premier Roy Romanow's deficit-fighting brand of social democracy, had older, more cautious followers, especially from the Prairies and from private-sector unions. McDonough, who had been late getting into the race, had acquired the support of many party regulars who were interested in keeping the faithful together and were afraid that Robinson's radicalism might split the party. She had party heavyweights like Gary Doer and Bob Rae in her corner, as well as many less well-known but experienced party workers.

The events on Saturday, 14 October, were little short of astonishing. On the first and only ballot, Robinson led with 655 votes, McDonough ran second with 566, and Nystrom was third with 514. Delegates started to prepare for a run-off between the two top contenders as Nystrom indicated his support for McDonough. But before the second ballot could be held, Svend Robinson crossed the floor and told McDonough that he was going to withdraw, allowing her to become leader by acclamation. It was an unprecedented move; no front runner had ever withdrawn before the final ballot in a Canadian nominating convention.

What can rational-choice analysis contribute to understanding these events? Let us look first at Robinson's decision. There were 1735 votes cast on the first ballot, so a majority would be 868. Robinson, at 655 first-ballot votes, was 213 away from that magic number. Ignoring the small numbers who might not vote on the second ballot or who might defect from his or McDonough's camps, he would need to attract those additional 213 votes from the 514 Nystrom supporters; that is, about 42 per cent of Nystrom's supporters would have to move to Robinson on the second ballot. Alan Whitehorn, a political scientist who has written more than anyone else about the NDP, was at the convention, and he is convinced that Robinson could not have attracted that many of Nystrom's supporters and hence had no realistic chance to win on the second ballot.[14] If Robinson shared that assessment, withdrawing after the first ballot was like a chess player resigning when checkmate becomes inevitable; there was no point in playing out an obvious endgame. Moreover, Robinson apparently thought his magnanimous gesture would help unify the party, though it also disappointed some of his enthusiastic supporters.[15]

The other surprise of the day was that McDonough, not Nystrom, made it to the second ballot, even though McDonough had been far behind in the primary voting. But the primary results were not binding, and the party activists who typically attend conventions would not have been altogether typical of the rank-and-file members who voted in the primaries. Moreover, since delegates cast secret ballots at Canadian nominating conventions, it is always difficult to know in advance what will happen. Nothing said or done before the ballot can be interpreted as binding.

Although McDonough was widely considered to be running third before the convention began, there is evidence that she had momentum. Her organization had been collecting endorsements from influential people in the party, she gave a well-received speech on Thursday night, and her people staged a noisy floor demonstration on Friday when she was officially nominated.[16] One explanation of her surprising second-place finish is simply that she peaked at the right time and attracted the support of many uncommitted delegates or lukewarm Nystrom supporters.

However, it is also possible that *strategic voting* played a part in her victory. In rational-choice parlance, 'voting directly in accordance with one's preferences' is sincere voting, while strategic voting is 'voting that is not sincere and that is intended to bring about preferred outcomes.'[17] In the run-off system, where voting means choosing one name on each ballot, strategic voting means picking the name of someone who is not your first choice at that stage.

We know from surveys of delegates at several nominating conventions that as many as 25 per cent of delegates may vote strategically on the first ballot. Typically, such voters favour one of the acknowledged

116   Game Theory and Canadian Politics

leaders but wish to vote on the first round to honour a personal friend, a regional favourite, or a long-serving party loyalist. They may think that such an indication of support will help that person get a cabinet seat or other high position, such as deputy leader, or they may wish to prevent the embarrassment of a very low total. However, this sort of behaviour constitutes strategic voting only in a loose sense. Those who indulge in it don't really wish to affect the final outcome of the convention. Believing that more than one ballot will be required, they are merely playing for time, serving a subsidiary objective before getting down to the serious business of electing the leader. The surveys show clearly that these strategic voters return almost 100 per cent to their true first choice as the voting progresses.[18]

In contrast, strategic voting in a stricter sense would take place on the first ballot if delegates were to vote for their second choice, rather than their first choice, to ensure that their third choice would be defeated. For an example, assume that the profile of support in Table 7.1 refers to a Canadian nominating convention. If all delegates vote sincerely on each ballot, candidate b will be eliminated on the first ballot and candidate a will win on the second ballot because the supporters of b will vote for a, their second choice, after b is eliminated. But this outcome is vulnerable to strategic voting by members of Group 2, whose first choice is c but whose second choice is b. As rational players, the members of Group 2 will see that their favourite, c, is doomed to lose and they will end up with their third choice, a, if they vote sincerely. But they will also see that, if they vote strategically for b on the first round, c will be dropped from the ballot and b will go on to win a majority in the second and final ballot. Thus they can guarantee a victory for their second choice, b.

According to Hugh Winsor, a senior journalist for the *Globe and Mail*, this is exactly what happened at the NDP convention. Based on an interview with McDonough's floor manager, Winsor reported the occurrence of

high-risk strategic voting intended to deny Mr. Robinson the leadership. Specifically, the outcome turned on the votes cast by 20 to 40 supporters of former Saskatchewan MP Lorne Nystrom who succumbed to the pitch of high-powered McDonough floor organizers (including Ontario NDP Leader Bob Rae). Their pitch was that the only way to stop Mr. Robinson was to boost Ms. McDonough to second place and thus onto the final ballot. Some of the converts continued to wear their Lorne buttons while they marked their ballots for Ms. McDonough.[19]

The 'Right Stuff': Choosing a Party Leader 117

This account is plausible. McDonough led Nystrom 566 to 514 on the first count. If only 27 of the McDonough voters were strategic voters who actually preferred Nystrom, and if these 27 had voted for their true first choice, Nystrom would have finished ahead of McDonough, 541 to 539. It is certainly within the realm of possibility that McDonough's organizers moved at least 27 Nystrom supporters to vote strategically by persuading them that Robinson could win a run-off against Nystrom but not against McDonough. However, we do not know this for sure. As mentioned, the voting was by secret ballot, and no researcher conducted a survey of delegate opinions at this convention. We thus lack the kind of data collected by Perlin at the 1983 Conservative convention and Archer at the 1989 NDP convention.

Was McDonough a Condorcet winner? Probably, although again we cannot be absolutely certain. Assume that there was a minimal amount of strategic voting for McDonough by Nystrom supporters – 27 votes, to be exact. This means that her true first-preference count was 566 – 27 = 539, while Nystrom's true first-preference count was 514 + 27 = 541. Since an overall majority was 1735/2 = 868, she would have needed an additional 868 – 539 = 329 votes to defeat each of the others in a head-to-head contest. For her to get those extra 329 votes in the contest with Nystrom, she would have had to command 329/655 = 50.2 per cent of the second preferences of the Robinson supporters – a virtual certainty. To beat Robinson, she would have had to command 329/541 = 60.8 per cent of the second preferences of the Nystrom supporters – quite likely in view of Gary Doer's statement that she was 'everybody's second choice' and in view of Svend Robinson's withdrawal from the race, which was based on the assumption that most Nystrom supporters had McDonough as a second choice.

In my view, there probably was a degree of strategic voting; and if there was, that made it possible for the run-off procedure to select the 'right' candidate, namely the Condorcet winner. Such a conclusion is a two-edged sword. On one side, it shows that the run-off procedure is in fact capable of choosing the 'wrong' candidate, and this aspect has to count as a criticism against it. But on the other side, it shows that delegates, realizing how the run-off works, can take control through strategic voting to make the system produce the 'right' result.

In a wider perspective, the evidence is also mixed. In 1983 (probably) and in 1989 (almost certainly), the 'right' candidate won on the last ballot; but the run-off eliminated candidates (Crosbie and Langdon) who would have beaten the eventual second-place finishers (Clark and Barrett) in head-to-head contests. In 1995, the Condorcet winner might well have been eliminated on the first ballot except for strategic voting. There is thus considerable evidence that the run-off procedure, although it is supposed to build consensus, can operate perversely to eliminate candidates with wider support than those who survive. Yet, disaster has been avoided thus far; at the end of the day, the 'right' candidate won in each case, and that is all that really matters.

Readers will have to make up their own mind about the merits of the Canadian run-off system; personally, I can't think of anything better for a nominating convention. It has emerged through trial and error, and the details have been honed through the practical experience of several parties. It is demonstrably efficient in the sense that, with n candidates, it produces a winner in at most $n - 1$ ballots – no small consideration when delegates have to get home and the convention centre may be booked for some other event the next day. Admittedly, it has the potential to produce a perverse result, but there is no evidence of that occurring very often. Moreover, when it does threaten to happen, the delegates can see it happening and may be able to forestall it by strategic voting, as they did in the case of Alexa McDonough.

In short, we can consider the Canadian-style run-off procedure an imperfect but reasonably acceptable way for a party to pick the 'right' leader, as long as we remember that the word 'right' has a limited meaning in this connection. It does not mean the wisest or most electable; it simply means the person most broadly acceptable to the delegates gathered at a particular convention. Whether their choice is 'right' in a larger sense has to be determined by subsequent events.

In a medium-term perspective, Brian Mulroney was certainly the right leader for the Conservatives. He led them to two majority governments and nine years of power. More particularly, he made the breakthrough among French voters in Quebec that had eluded his immediate predecessors, Robert Stanfield and Joe Clark. Yet in a longer-term perspective, Mulroney's high-risk strategy of reopening the Constitution led to the break-up of the federal Conservative coalition, as Reformers defected in the West and the Bloc Québécois was formed in Quebec, and the Conservatives were reduced to two seats in the general election of 1993. Historians will argue forever over whether he had the 'right stuff.'

Judgments on the NDP leaders seem easier to make. In 1988, the NDP, under the leadership of Ed Broadbent, had its best election result ever – 43 seats and 20 per cent of the popular vote. Audrey McLaughlin got some 'bounce' from her convention victory and endless publicity for being the first woman to lead a major federal party. For about a year after she was selected leader, the NDP was the most popular federal party, topping out in the polls at 41 per cent in January 1991. But then an inexorable decline set in. The NDP fell to single-digit levels (8 per cent) by July 1993 and never recovered. In the general election of October 1993, the party won just nine seats and 6.9 per cent of the popular vote – its worst showing since John Diefenbaker almost annihilated the CCF in 1958.[20] Audrey McLaughlin may have been a Condorcet winner at the convention, but her leadership was a disaster for her party.

There has not been time to make a final judgment about Alexa McDonough, but she did better than expected in the 1997 federal election, when the NDP regained official-party status in the House of Commons with 21 seats. Drawing on her local reputation, she led her party to victory in 13 ridings in Atlantic Canada – an unprecedented breakthrough in a region where the NDP had always been marginal. But the NDP is still far from the total of around 40 seats that it enjoyed in the 1980s, and only time will tell whether McDonough can take her party back to that level. It certainly will not be easy as long as the Reform Party continues to dominate the West, blocking the NDP out of many traditional strongholds in Saskatchewan and British Columbia.

Perhaps the moral of the story is that rules for choosing a leader can only aggregate the views of party members. The rules are important; they can function more or less efficiently, and sometimes perversely; but they cannot work miracles. If the members of a party are ideologically out of step with their competitive environment, they cannot save their party by tinkering with leadership selection rules.

This is not to say, however, that bigger changes in the selection procedure are unimportant. In recent years, there has been a strong trend toward choosing party leaders through some sort of direct election by the membership rather than at a nominating convention.[21] Techniques have ranged from mail ballots to drop-off ballots to telephone voting, but these variations are secondary compared to the big difference between direct election and a convention. Direct election empowers the party's grassroots membership, whereas a convention tends to be dominated by party loyalists, workers, and insiders. It is striking that Ontario's Mike Harris and Alberta's Ralph Klein were both chosen leader by direct election, apparently contrary to the desire of the Progressive Conservative party establishment in both provinces, and both went on to win elections and become premiers. Changes of this type, which affect the entire process of running for leader, are more important than narrower changes in the decision rules.

8

The Staying Power of the Status Quo[1]

Introduction: The *Morgentaler* Decision

In its *Morgentaler* decision of January 1988, the Supreme Court of Canada struck down the centrepiece of Canada's abortion law, section 251 of the Criminal Code. The court held that it conflicted with section 7 of the Charter of Rights and Freedoms, which guarantees the 'security of the person.' This ruling radically changed the legal status quo; it was as if Parliament had repealed that section of the Criminal Code without passing a replacement amendment.

However, the justices in their opinions invited Parliament to try its hand at new legislation. Justice Bertha Wilson hinted broadly that a gestational approach, with differential rules based on fetal stage of development, would pass constitutional muster: 'The precise point in the development of the fetus at which the state's interest in its protection becomes "compelling" I leave to the informed judgment of the legislature which is in a position to receive guidance on the subject from all the relevant disciplines.'[2] Similarly, Justice Jean Beetz, referring to foreign jurisdictions where the gestational approach had been legislated, wrote: 'It is possible that a future enactment by Parliament along the lines of the laws adopted in these jurisdictions could achieve a proportionality which is acceptable under s. 1 [of the Charter].'[3]

Even apart from these broad hints about the gestational approach, the Supreme Court's decision left the door open for legislation. Six of the seven members who participated in *Morgentaler* held that there was no right to an abortion under the Charter and that the state had an interest in protecting the unborn child. Of these six, the four who ruled against section 251 did so because of the unfairness and delays caused by its procedural complexity. A simpler abortion statute, relying, say, on the opinion of two physicians rather than on a therapeutic-abortion committee, might have had a reasonable chance of surviving the tests imposed in *Morgentaler*.

Although the Court virtually invited Parliament to pass new legislation and the government of Brian Mulroney announced its determination to do so, no new legislation was ever passed, even though Mulroney's government had an overwhelming majority in the House of Commons before the election of November 1988 and a still comfortable majority afterward. The first of two attempts ended in failure, when all options were defeated in a series of free votes in the House of Commons; the second passed the House of Commons but failed on a tie vote in the Senate.

The failure to pass new legislation illustrates two principles: (1) the staying power of the status quo when there is a cyclical opinion structure in the legislature, and (2) the influence of parliamentary procedures upon legislative outcomes. This chapter draws on the theory of voting games to explore both these principles.

Rational Choice and Parliamentary Procedure

Literature on the aftermath of *Morgentaler* has emphasized the fragmentation of opinion in Parliament, which produced paradoxical coalitions able to block all proposals but not able to pass anything.[4] When there are various options and multiple dimensions, cycling is common; that is, any outcome can be overturned by at least one other outcome, which in turn can be overturned by another, and so on, as illustrated in the simplest case by the paradox of voting.[5] The preceding chapter presented a purely numerical example of the paradox of voting; here is another example, with a few hypothetical facts attached to it.

Imagine that a committee grappling with the issue of pornography is divided into three factions of roughly equal size: conservatives, feminists, and libertarians. The conservative theory is that all pornography should be restricted because it is immoral. The feminist theory distinguishes between pornography, which should be outlawed because it degrades women, and erotica, which should be legalized because it simply portrays sexuality. The libertarian theory is that all pornography should be unrestricted because, as long as the viewing is voluntary, it does not infringe anyone's rights.

Assume that the conservatives propose a bill to prohibit all pornogra-

TABLE 8.1
Cyclical Preference Structure on Pornography

Faction	Preference Ordering		
	First	Second	Third
Conservatives	c >	l >	f
Feminists	f >	c >	l
Libertarians	l >	f >	c

phy (option c). The feminists react with an amendment to outlaw only material that degrades women (option f). For their part, the libertarians wish to have no legislation at all (option l). Assume that the three factions rank these options as shown in Table 8.1.[6]

Remember the concept of the Condorcet winner, defined as an option that 'can defeat all others in a series of pairwise contests.'[7] This preference structure has no Condorcet winner. In a choice between c and l, c will win because it is preferred by both the conservatives and the feminists. But f, supported by a coalition of feminists and libertarians, will defeat c; and l, supported by a coalition of libertarians and feminists, will defeat f. So $c > l > f > c$ – a decisional cycle in which any option can be beaten by some other option. As in all such cycles, the aggregation of individually rational preference orderings gives rise to a collectively intransitive preference structure.

The actual result will depend crucially on the order in which the options are voted upon. If, for example, the conservatives begin by moving their preferred policy, c, they can be defeated by the feminists moving f as an amendment (the feminists and libertarians prefer f over c); but then f will be defeated by l in the next round (the conservatives and libertarians prefer l over f). Any outcome is possible, depending on the order of motion and amendment, and any outcome can be overturned by another option if the rules permit reopening the question after a series of votes.

The importance of order of voting in this illustration points to the crucial role of parliamentary decision-making rules, particularly the so-called amendment procedure, which requires amendments to be voted on serially until the final vote, in which the choice is between the final version of the bill and no legislation at all (effectively the status quo). In the British tradition, the normal procedure is for a bill to be introduced at first reading, sent to committee at second reading, and then brought back to the floor of the chamber, after which amendments can be moved and voted upon one at a time. At each such vote, the choice is between the original bill and the bill as it would be affected by the amendment; there is as yet no opportunity to vote on the status quo. That opportunity only arrives at the last stage. After all amendments have been disposed of, there is a vote on the bill in final form at third reading, which in effect pits the bill against the legal status quo. If the bill is defeated at that stage, the status quo prevails; if the bill is passed, the status quo is overturned.

The great merit of the parliamentary amendment procedure is to prevent cycling. By requiring a series of pairwise choices in a stipulated order, it ensures that there will be a definite outcome supported by a majority vote. It also ensures that the Condorcet winner (if there is one) will be chosen. By definition, a Condorcet winner defeats every other option, so it will survive every vote in a multistage process to emerge as the final winner. If there are only three choices – the original motion, an amendment, and the status quo – the amendment procedure also guarantees that the status quo will triumph if neither the motion nor the amendment is a Condorcet winner. If the status quo is a Condorcet winner, it wins automatically in the second round. If there is no Condorcet winner, then both the motion and the amendment win against one choice and lose against another. Thus, no matter whether the amendment beats the motion or the motion beats the amendment, the winner of the first round must lose to the status quo in the second round.[8]

Things are more complicated when there are more than three options on the table. A Condorcet winner will always triumph; but, in the absence of a Condorcet winner, the amendment procedure may allow an alternative (a) to win even though the legislators would prefer another (b) over a. This scenario can happen if b is defeated in an earlier vote against c, which then loses to a. Option b would defeat a if they could go head to head, but that pairing does not occur. As a general rule, the amendment procedure privileges the legal status quo because it is not voted upon until the end, after most other options – including some that might have beaten the status quo – have already been eliminated. This nonobvious conclusion about the importance of procedural rules illustrates the usefulness of the rational-choice approach in exploring the workings of political structures and processes.

Game theorists agree that cycles based on diversity of preferences are extremely common and that equilibrium, where it exists, is usually induced by political institutions, including decision-making processes such as the amendment procedure. Political institutions act like filters for removing options and like funnels for directing choices into binary alternatives and single dimensions so that equilibria nominally supported by majorities can emerge. Theorists such as Shepsle, Fiorina, Riker, and Jones have called this phenomenon *structure-induced equilibrium*.[9] In effect, this chapter is a case study of structure-induced equilibrium in the Canadian Parliament.

Free-Vote Failure

After the Supreme Court released its decision, Prime Minister Mulroney quickly announced that his government would introduce a bill, but the Conservative caucus proved deeply divided on the issue. Even after five meetings, no consensus emerged. Mulroney then announced that he would combine a free vote in the House of Commons with a unique procedural initiative. The government would introduce not a bill, but a set of resolutions designed to test opinion in the House. The main resolution would take a gestational approach to abortion. In 'the earlier stages of pregnancy [undefined],' abortion would be allowed on the opinion of a single medical practitioner 'that the continuation of the pregnancy of a woman would, or would be likely to, threaten her physical or mental well-being.' In 'the subsequent stages of pregnancy [also undefined],' abortion would require the opinion of two practitioners that the pregnancy 'would, or would be likely to, endanger the woman's life or seriously endanger her health.'[10] This main resolution was accompanied by two amendments, both of which would have changed it substantially by abandoning the gestational approach. The first was a pro-life amendment restricting abortion at all times to cases of danger to life or health verified by two physicians. The second was a pro-choice amendment allowing abortion at any point based on the decision between a woman and her doctor. Together, the resolution and its amendments constituted a poll on a set of three options – pro-life, pro-choice, and a gestational compromise. Lurking behind the explicit options was an implicit alternative – the new, judicially created status quo of no law at all.

According to Deputy House Leader Doug Lewis, who introduced the resolutions, the government would proceed to draft a bill based on whichever option received the most support in the House.[11] This procedural manoeuvre was designed to deliver the Conservative caucus from its internal stalemate between pro-life and pro-choice factions by producing a House of Commons plurality, maybe even a majority, in favour of the gestational compromise. The pro-choice option, voted upon first, would almost surely have been defeated. The pro-life option, voted upon second, would also probably, though less certainly, have failed. With both 'extremes' defeated, the gestational compromise would then presumably have looked more appealing on the final vote. The strategy was not foolproof, because it was also open to members to defeat the main resolution; but even if it had failed to get a majority, it might well have received more votes than either of the amendments. If it had been allowed to run its course, the process might have been psychologically effective in steering MPs, particularly in the Tory caucus, to accept the gestational compromise.

From the standpoint of rational choice, the procedure was an attempt to get around the privileged position of the status quo. It amounted to saying to Parliament: 'Something has to be done [the prime minister's premise] and here are three options – pick one.' Indeed, the pro-choice option offered as a vote was in fact scarcely different from the post-*Morgentaler* status quo; so by asking members to vote on it first, the government was in effect trying to get them to agree at the outset that the status quo was unacceptable. It was thus reversing the normal parliamentary procedure, in which the status quo does not appear until the last vote as the implicit alternative to the bill as finally amended.

All of this was an inspired example of what Riker has called *heresthetic*, the manipulation of decision-making procedures to achieve particular results.[12] But unfortunately for the prime minister, the opposition parties would not accept his new rules. First they refused unanimous consent, causing the government to withdraw its package of resolutions; then they appealed to the speaker when the government reintroduced its motions.[13]

After this setback, the government moderated its attempts to revise Commons procedures; but it continued with its broader strategy of testing the mood of the House through holding a free vote on a resolution rather than by introducing a bill. On 26 July 1988, Doug Lewis introduced a motion containing the gestational compromise without the pro-life and pro-choice amendment options.[14] MPs would now be free to introduce their own amendments according to ordinary parliamentary procedure.

Seventy-five MPs spoke in a prolonged debate that started on 26 July and carried on into the early morning hours of 29 July. They introduced 21 amendments, of which the speaker disallowed all but five on grounds of being redundant or contrary to the sense of the resolution. The five that came to a vote (all were defeated) can be described as follows:[15]

One: A pro-choice amendment by Mary Collins (PC, Capilano-Howe Sound) that would have removed the need to show any grounds for abortion in the earlier stages of pregnancy and would have defined the mother's health in the later stages as being both 'mental and physical.' Actually, Collins was satisfied with the legal status quo created by *Morgentaler* but was trying to head off a pro-life victory by giving the pro-choice forces something they might be able to support.[16] However, her motion was defeated 191–29. Eighteen of the 29 who voted for the Collins amendment would ultimately support the government's resolution but were willing to make this concession to the pro-choice side.

Two: A pro-life amendment by Ken James (PC, Sarnia-Lambton) that would have defined the 'earlier stages of pregnancy' as the first trimester. This motion was intended to assuage the fear of many pro-lifers that the yet-to-be-drafted legislation would define 'earlier stages' so expansively that hardly any abortions would be prohibited. James's motion was defeated 202–17. Most of the 17 who supported it would ultimately vote for the government's resolution but were willing to make this concession to the pro-life side.

Three: A far stronger pro-life amendment by Gus Mitges (PC, Bruce-Grey) that would have abandoned the gestational approach and restricted abortion to circumstances that threatened the mother's life. This motion failed narrowly, 118–105.

Four: An amendment by Bobbie Sparrow (PC, Calgary Southwest) and seconded by Ken James to define the 'earlier stages of pregnancy' as the first 18 weeks. This motion was defeated on an unrecorded vote.

Five: A strong pro-choice amendment by John Bosley (PC, Don Valley West) to remove all conditions for abortion except that it be performed by a 'qualified medical practitioner.' In practice, this amendment would have endorsed the legal status quo set by *Morgentaler*. But it would have left intact the preamble of Lewis's resolution, which spoke about balancing 'the right of a woman to liberty and security of her person and the responsibility of society to protect the unborn,' and those words might have been invoked in future litigation. The Bosley motion was defeated 198–20. Almost all of the 20 who voted for it had also supported the pro-choice Collins amendment.

After all these amendments had failed, the government's gestational approach was put to the test, and it too lost, 147–76, defeated by a paradoxical coalition of strong pro-choice advocates who wanted no legislation and had voted against all the amendments (except for a few who had supported the Collins and/or Bosley amendments), and of strong pro-lifers who had voted for the Mitges amendment. 'Les extrêmes se touchent,' as the old French proverb says. As a result, nothing passed, and the post-*Morgentaler* legal status quo continued to prevail.

As shown in Figure 8.1, it is possible to arrange the 224 MPs who voted on these motions along a single dimension ranging from strong pro-choice to strong pro-life.[17]

Since a single dimension is clearly at work here, why did a majority not coalesce around the position of the median voter in the House of Commons, as is predicted by elementary rational-choice analysis of unidimensional conflicts?[18] In this array of 224 MPs, the median voter (between the 112th and 113th positions) would have been located among the 'Pro-Life–Leaning Moderates' who supported the James amendment as well as the government's resolution. However, the vote on the James amendment came too early, while the 96 'Resolute Pro-Lifers' and nine 'Pro-Life Compromisers' were still hopeful of passing the

FIGURE 8.1
One-Dimensional Arrangement of Voting Blocs

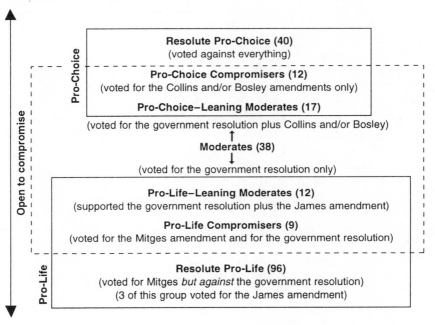

Mitges amendment. But why did so many pro-lifers refuse to vote for the gestational compromise *after* they saw own first preference, the Mitges amendment, defeated? Nine Mitges supporters (the 'Pro-Life Compromisers') did swing to the government resolution in the end, but 96 did not. Were they acting irrationally? Did they not realize that by helping to defeat the gestational compromise they were perpetuating the legal status quo?

To answer such questions, one must recall the exceptional nature of these votes. Members were not voting on a bill, but on a resolution and ancillary amendments intended to guide the drafting of legislation not yet introduced. It was never stated that only an option supported by a majority would be incorporated into the future bill. Thus Mitges could claim the endorsement of a plurality after all options had failed to get a majority: 'It is a victory,' he said, 'because my motion got the most votes.'[19] For committed pro-lifers, the situation amounted to one of strategic voting to defeat the government's resolution in this round in hopes of getting a better deal from the cabinet in the next round. The incentive to vote strategically transformed the situation from a unidimensional conflict with a potential Condorcet winner at the median voter to a cycle with no winner except the one induced by the decision-making rules.

However, the pro-life forces turned out to be mistaken in their estimate of future probabilities. Not only did they prove unable to get a better deal than the gestational compromise, but they never succeeded in getting anything at all passed in Parliament as described in detail in the next two sections of this chapter, 'Second Try' and 'Senate Stalemate.' This miscalculation by the pro-life forces delivered to the pro-choice forces a victory they could never have secured on their own. There were, on a maximum count, only 69 pro-choicers voting on the issue (40 'Resolute Pro-Choice' plus 12 'Pro-Choice Compromisers' plus 17 'Pro-Choice–Leaning Moderates'). The pro-lifers did not take full account of the staying power of the status quo.

To see the effect in this case, consider what might have happened if parliamentary procedure called for an initial vote on whether the status quo was acceptable. To proceed in this way is not always irrational; we sometimes make major decisions without knowing what the alternative will be. Nations, for example, often go to war without being certain that they can win. Declaring war in such circumstances amounts to saying that the status quo, even though it is peaceful, is worse than any likely alternative.

Staying Power of the Status Quo 129

So imagine the House had begun by voting on whether to pass any legislation at all. On that question, if all moderates and pro-lifers had voted yes, the judicially created status quo would have been rejected 172–52, or 155–69 if the 'Pro-Choice–Leaning Moderates' had voted for the status quo. With the status quo removed from further consideration, the choice would then have lain between the gestational compromise and the Mitges amendment. At that stage, the pro-choice forces would presumably have supported the government resolution because it was closer to their ideal point than was the Mitges amendment; and the government resolution would have carried 110–105 (in this scenario, the nine Mitges supporters who voted for the government resolution after the defeat of their first choice would have voted for the Mitges amendment).

A schematic representation of this scenario, omitting the minor amendments and considering only the main options – the status quo, the government resolution, and the Mitges amendment – is shown in the upper panel of Figure 8.2 under the heading of 'hypothetical procedure.'

In the first stage, shown on the upper left, the pro-life and moderate factions coalesce to decide that legislation is necessary, thereby defeating the pro-choice bloc. In the second stage, shown on the upper right, the moderate and pro-choice factions combine to defeat the Mitges amendment and pass the government's resolution, thereby defeating the pro-life faction.

This scenario is labelled 'hypothetical' because it did not and could not take place. The actual course of events was determined by the standard amendment procedure shown in the lower panel of Figure 8.2. In the first stage, depicted on the lower left, the pro-choice and moderate factions combine to defeat the pro-life Mitges amendment. In the second stage, shown on the lower right, the pro-life and pro-choice factions combine to defeat the government's resolution, thereby leaving the post-*Morgentaler* legal status quo intact.

Figure 8.2 is an excellent example of a structure-induced equilibrium. When there is no natural equilibrium in the sense of a Condorcet winner, the rules of the game prevent cycling by determining an outcome. In this real-life instance, there were three main alternatives and no majority for any of them; so the conventions of parliamentary procedure brought about a result.

That opinion in the House of Commons was divided among pro-choice, pro-life, and moderate alternatives reflected the larger reality of

FIGURE 8.2
Voting Procedures in the House of Commons

Hypothetical Procedure

First Vote			Second Vote		
Alternatives	Coalitions	Winner	Alternatives	Coalitions	Winner
Status quo vs. Legislation	Pro-choice vs. Moderates + Pro-life	Legislation	Mitges vs. Government resolution	Pro-life vs. Moderates + Pro-choice	(Government resolution)

Standard Amendment Procedure

First Vote			Second Vote		
Alternatives	Coalitions	Winner	Alternatives	Coalitions	Winner
Mitges vs. Government resolution	Pro-life vs. Moderates + Pro-choice	Government resolution	Status quo vs Government resolution	Pro-life + Pro-choice vs. Moderates	(Status quo)

◯ = Structurally induced equilibrium

Canadian public opinion. Janine Brodie, herself an ardent supporter of the pro-choice side, has written:

> Canadian public opinion on abortion appears quite stable. Approximately one-quarter of Canadians agree that abortion should be legal under any circumstances, another 13% believe that it should be illegal under all circumstances, and the vast majority think abortion should be allowed in certain circumstances. Among this latter group, Canadians are more likely to support abortion for the so-called 'hard cases' of rape, incest, the woman's health, and fetal deformity and less likely to support 'soft cases' relating to socio-economic and life-style factors.[20]

In a Gallup poll taken in September 1988, a few weeks after the vote in the House of Commons, 20 per cent of respondents said that abortion should be 'legal under any circumstances,' 13 per cent said that it should be 'illegal in all circumstances,' and 65 per cent said that it should be 'legal only under certain circumstances.' Of this middle group, large majorities accepted Brodie's 'hard cases' – danger to health (89 per cent), rape or incest (76 per cent), and fetal defect (68 per cent) – as legitimate reasons; but only small minorities accepted her 'soft cases' as legitimate – agreement between the woman and doctor (32 per cent), termination within three months of conception (24 per cent), low family income (14 per cent), and termination within five months of conception (6 per cent).[21] Given this unimodel distribution of views, with relatively high agreement among the moderates about what circumstances justify abortion, it should have been possible for Parliament to craft a compromise position to command support among the middle majority of public opinion. That this did not happen shows that political developments are not inevitable, and that outcomes are conditioned by institutional decision-making rules and by strategic behaviour within those constraints.

Finally, it is worth asking whether the Mulroney government brought this failure upon itself by adopting poor strategy. Faced with a divided caucus, it tried and failed to overcome those divisions by finding the median voter in the larger forum of the House of Commons. That was one strategic choice, but there were other options. There was probably a potential majority within the Conservative caucus for a pro-life position. Out of 206 members in the caucus at that time, 87 voted for Mitges and against the government, while nine voted for Mitges and for the government after Mitges failed; and another 12, while not going so far as to support Mitges, voted for the James amendment, which was at least mildly pro-life in intent. If the prime minister had thrown his influence in with these 108 members, who were pro-life to varying degrees, he might have carried a number of other moderates with him to create a pro-life majority within caucus. Something like the Mitges amendment, perhaps tempered a bit, could then have been introduced into the House as a government bill that all Tories would be expected to support.

However, this scenario would have shattered cabinet solidarity by putting prominent pro-choice female ministers such as Pat Carney, Barbara McDougall, Flora MacDonald, and Mary Collins in an untenable position. Mulroney had gone out of his way to put women in the cabinet and advance their careers.[22] His sponsorship now gave them something like veto power on this issue, even though the pro-choice faction was smaller than the pro-life faction in caucus.

In any case, even if the prime minister had tilted pro-life and used his huge majority plus party discipline to force the bill through the House, he would probably have been blocked in the Senate. In July of 1988, the Liberals still had a majority in the Senate; and Liberal leader John Turner had asked them to use that majority to block the Conservative government's Bill C-130, the enabling legislation for the Free Trade Agreement.[23] Under those confrontational circumstances, there would have been little chance that the Liberal-dominated Senate would have quickly passed a controversial abortion bill drawn up by the Conservative government and whipped through the House of Commons under party discipline.

In the event, the prime minister, opting for an election, asked the governor general to dissolve Parliament on 1 October 1988. Even if it had passed the House, an abortion bill would likely have died on the Senate's order paper, given the usual lapse of time required to move legislation through that chamber. Brian Mulroney might conceivably have solved his collective-choice problem through party discipline, but he probably would have been undone by the other procedural roadblocks built into Canada's bicameral parliamentary system.

Second Try

In the fall of 1989, after a national election and further controversial judicial decisions on abortion, the Mulroney government again set the legislative wheels in motion. A Conservative caucus committee was appointed but failed to reach a compromise among its pro-life, pro-choice, and moderate factions. Then the cabinet adopted a new tack. Bill C-43, introduced into the House of Commons on 3 November 1989, abandoned the gestational approach in favour of a compromise now based on calculated ambiguity. The bill made abortion a criminal offence except when pregnancy threatened 'the life or health of the female person,' which seemed to go quite far in the pro-life direction; but it defined health as 'physical, mental and psychological health,' which seemed to lean in the pro-choice direction.[24]

This compromise, however, did not satisfy either side. The pro-life side feared that the word 'psychological' would be interpreted so broadly, encompassing all sorts of social and economic factors, that it would amount to abortion on demand, while the pro-choice side did not want any law that made abortion a criminal offence. Thus Bill C-43 seemed to face the same sort of paradoxical coalition that had defeated the 1988 resolution. A poll of 232 MPs published just before second reading, scheduled for 29 November 1989, found 127 opposed to the bill and only 105 in favour.[25]

The prime minister reacted to this threat of defeat with a combination of leadership and heresthetic manoeuvres. He made a speech in the House at second reading in favour of the bill, thus engaging his influence in a way that he had not done for the previous resolution. He also announced that all cabinet ministers – 40 at the time – would vote for the bill; the free vote would be for backbenchers only. More importantly in terms of numbers of votes affected, Mulroney made it clear that, if Bill C-43 was defeated, the government would not try again to bring in abortion legislation, at least not before another election.[26] This position undermined the rationale for strategic voting that had animated many of the 96 MPs who, after supporting the Mitges amendment, had voted against the government resolution in 1988, hoping they could get legislation more to their liking. To use the language of game theory, they had played Chicken with the prime minister. Now he was responding in an often-recommended fashion for playing Chicken – precommitment.[27] If he could make his backbenchers believe that this truly was their last chance to secure any kind of legislation, it would be rational for them to accept the compromise embodied in this piece of legislation.

Figure 8.3 models the situation as a pair of sequential games between the pro-life bloc and the prime minister. The bloc's first preference is for another, tougher bill; its second choice is for the present bill; and its last choice is for no bill at all. Since the game is sequential, the bloc must decide how to vote based on what it thinks the prime minister's preferences are. If it believes Scenario A represents his preferences, it should defeat the bill; he will then bring in another bill more to its liking, because no bill at all is his last choice. But if the bloc believes Scenario B represents the prime minister's thinking, it should support the present bill because he will not introduce another one if this one is defeated. The best the bloc can do in Scenario B is to settle for their second choice (Bill C-43) rather than its third choice (no bill at all).

By precommitting himself with his speech at second reading, Mulroney apparently succeeded in making most of the pro-life bloc accept Scenario B. Of the 96 MPs (89 of them Conservatives) who had supported the Mitges amendment and then rejected the government's resolution, 49 had been returned to the House after the 1988 election. Although diminished in numbers, they would still have been able to block legisla-

134 Game Theory and Canadian Politics

FIGURE 8.3
Sequential Games between the Pro-Life Bloc and the Prime Minister

Scenario A: Before Precommitment

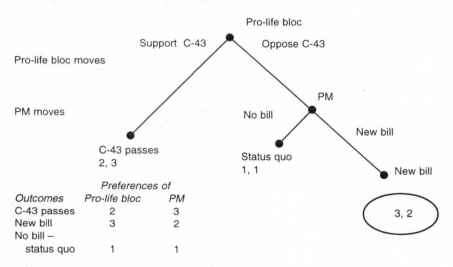

Outcomes	Preferences of Pro-life bloc	PM
C-43 passes	2	3
New bill	3	2
No bill – status quo	1	1

Note: 3 is preferred over 2, which is preferred over 1

Scenario B: After Precommitment

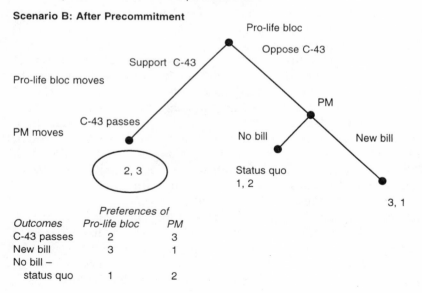

Outcomes	Preferences of Pro-life bloc	PM
C-43 passes	2	3
New bill	3	1
No bill – status quo	1	2

Note: 3 is preferred over 2, which is preferred over 1

tion if they maintained their opposition to any compromise and voted with the pro-choicers. But this time, 32 of the 49 pro-lifers voted for C-43 on third reading, and only 10 opposed it.[28] Most had apparently come to the conclusion that it really was their last chance to get anything at all.

Precommitment, together with the other measures that Mulroney took, was sufficient to get Bill C-43 through the House, albeit just barely. It passed second reading 164–114 and third reading 140–131. Breaking the solidarity of the pro-life bloc and gathering up the smaller number of pro-choice Tories were both crucial to this narrow victory. In the end, while some pro-lifers from all parties continued to oppose it, the main resistance to C-43 came from the pro-choice forces among the NDP and Liberals who preferred to have no legislation at all.

Senate Stalemate

From the House of Commons, the action now shifted to the Senate, which ultimately produced its own extraordinary illustration of the staying power of the status quo. Bill C-43 received first reading on 30 May and second reading on 26 June 1990. At second reading it was referred to the Legal and Constitutional Affairs Committee, which tabled its report on 23 January 1991. Although the committee recommended no amendments, the extensive hearings that it had held provided debating points for both pro-choice and pro-life advocates. After a week of off-and-on debate, the final vote on third reading was held 31 January 1991.

Because of the prime minister's appointment of eight additional senators in 1990 to get the GST passed, the Conservatives now had a majority of seats in the Senate. By controlling the agenda and the committees, they could prevent the Liberals – who had tied up earlier Conservative legislation – from obstructing C-43. However, this was to be a free vote, so the Conservatives could not guarantee a victory through party discipline. In the event, the bill failed to get a majority since the senators split mostly, but not entirely, along party lines, as shown in Table 8.2.

Of the seven Conservatives who voted against the bill, three women recently appointed by Brian Mulroney were clearly pro-choice. Pat Carney did not speak to C-43 in the Senate; but when she had been in the Commons in 1988, she had supported the two pro-choice amendments to the government resolution. Conservative Senators Mira Spivak

TABLE 8.2
Senate Votes on Bill C-43, by Party

	Conservative	Liberal	Reform	Independent	Total
Yea	40	2	1	0	43
Nay	7	35	0	1	43
Total	47	37	1	1	86

and Janis Johnson did speak to C-43, and both openly opposed it on pro-choice grounds.

Of the Liberals who opposed C-43, many were also pro-choice, as shown in the remarks of Liberal Senators Royce Frith, Joyce Fairbairn, Michael Kirby, and Joan Neiman.[29] However, there was also an active group of Roman Catholic pro-life Liberals, spearheaded by Stanley Haidasz, who unsuccessfully introduced seven pro-life amendments. Liberal senators who either voted for at least one of these amendments or expressed pro-life views in the debate included Rhéal Bélisle, Peter Bosa, Ray Perrault, L. Norbert Thériault, Bernard Alasdair Graham, Gildas Molgat, Thomas-Henri Lefebvre, and of course Haidasz himself.[30] To this group must be added the Conservative John Michael Macdonald, who spoke against C-43 on pro-life grounds and also voted for all the Haidasz amendments.[31] Thus, at least nine senators felt that government's bill did not go far enough in the pro-life direction.

Those who defended C-43 in the Senate debates admitted that it was a compromise; William Doody, who introduced it to the floor, stated at the outset that it would not be 'satisfactory to persons who have strong personal convictions at either end of the spectrum of opinion on abortion.'[32] Several supporters, both pro-choice and pro-life, also admitted that they were not satisfied with the bill and were supporting it only because, in the words of William Kelly, it was 'better than nothing.'[33]

An oddity in the Senate proceedings was that the bill was defeated on a tie vote – reportedly the first time in the history of the Senate that this had happened.[34] In the House of Commons, this outcome could not have occurred; the speaker would have broken the tie with a so-called 'casting vote.'[35] But the Senate has a different arrangement, modelled after the British House of Lords.[36] The speaker of the Senate, appointed by the governor general and removable at any time, is a more partisan figure than the speaker of the House of Commons, which helps to explain the Senate's different procedure: 'Questions arising in the Senate shall be decided by a Majority of Voices, and the speaker shall in all Cases have a Vote, and when the Voices are equal the Decision shall be deemed to be in the Negative.'[37] In practice, Senate speakers act as impartial presiding officers most of the time; but they do on occasion participate in debate and cast votes.[38]

When Bill C-43 went through report stage and third reading, Speaker *pro tempore* Rhéal Bélisle was in the chair because the appointed speaker, Guy Charbonneau, was under investigation on a conflict-of-interest matter. After Senator Bélisle chose not to participate in the division on third reading, it would have been improper for him to try to vote at the end once the tie was announced. In any case, it was unnecessary; for he was already on record as opposing the bill on pro-life grounds and hoping for its defeat. Thus it would be misleading to say that C-43 was defeated in the Senate on a technicality. There was a majority against it in the chamber that day because the pro-choice and pro-life forces combined to oppose it, even if Senator Bélisle, who was aligned with the pro-life side, did not cast a vote.

It should be noted in passing that there are sound rational-choice reasons why legislation should require at least 50 per cent + 1 of votes to pass. The underlying reason for this rule is to prevent disequilibrium and cycling. If a motion could be passed on a bare 50 per cent, it could be immediately overturned by a contrary motion that would also get 50 per cent, and there would be no determinate outcome. Since the status quo is always the default option in the final vote on the motion, the status quo continues to prevail unless there is a clear majority to overturn it.

Broadly speaking, the configuration of opinion in the Senate was similar to that which had defeated the government's first attempt in the House of Commons. The bill was a compromise with which some of its own adherents were barely satisfied, and poised against it was a paradoxical coalition of pro-choice and pro-life advocates. Motivating the pro-life forces was a faith that defeating this proposal in the present might bring about better legislation in the future; as Ray Perrault said, 'Better the status quo, perhaps, and a return to the drafting process than a halfway measure.'[39] The chief difference from 1988 was that in 1991 the pro-life opponents of the bill in the Senate were mostly Liberals, whereas in 1988 in the House of Commons they had mostly been Conservatives.

As in 1988, parliamentary procedure also affected the outcome. The status quo prevailed because, as the default option, it was not voted upon until the end. If the Senate had voted first upon whether to accept or reject the status quo, the nine who opposed C-43 because it did not go far enough would logically have combined with the 43 senators who voted for the bill to support the proposition that the status quo was unacceptable and some legislation was necessary. Then, on the second and final vote, the moderates could have combined with the pro-choice faction to defeat a pro-life amendment and pass the government's bill.

Finally, it should also be noted that the very existence of a bicameral legislature is itself another defence of the status quo. If legislation must be passed in two houses rather than one, there are more obstacles in the path of success. Implied in Sir John A. Macdonald's famous characterization of the Senate as the chamber of 'sober second thought' is that bicameralism will cause the failure of some bills that otherwise would pass. In that sense, contemporary rational-choice analysis is an extension of older insights into the dynamics of parliamentary procedure.

At every attempt to legislate after the *Morgentaler* decision, the Mulroney government faced the same strategic dilemma. Members of both the House of Commons and the Senate were fragmented into three major blocs – pro-choice, compromise, and pro-life – none of which constituted a majority. The unwillingness of the extreme blocs, particularly the pro-lifers, to support a compromise turned the situation into a cycle in which any option could be defeated by a coalition of two out of the three blocs. But the rules under which legislatures operate do not permit cycling; they force a determinate outcome – a structure-induced equilibrium – by specifying a series of binary choices in a particular order. In the absence of a majority or a willingness to compromise around the position of the median voter, these rules favour the status quo, which in this case was an absence of legislation arising from the Supreme Court's ruling in *Morgentaler*; hence, the ironic outcome that nothing could get passed, even though majorities in both chambers disliked the legal status quo and wished to legislate. Because those who favoured legislation could not agree on its substance, the victory went to the minority who opposed any legislative initiative. The Supreme Court had given them the procedural high ground of the status quo; and, though outnumbered, they were able to defend it against the divided attackers.

This one case study illustrates an important contemporary trend in Canadian politics. Because it is so difficult to overturn the status quo in Parliament, interest groups wishing to change public policy are increasingly turning to the courts to get results. Litigation is not easy; on the contrary, it is both time-consuming and expensive. But a victory in the Supreme Court of Canada, or even in one of the courts of appeal, may produce a sudden reversal of the status quo that an interest group could never have achieved in Parliament. And once obtained, that victory will be difficult to roll back in Parliament, for the same reason that it would have been hard to achieve there: the staying power of the status quo.

9

Invasion from the Right: The Reform Party in the 1993 Election

Up to this point, the models presented in this book have dealt with situations of *discrete choice*. Such situations involve making the best choice among a small number of clearly distinct strategies: cooperate or defect; bat left or right; use metric or imperial; vote for Nystrom, McDonough, or Robinson; and so on. But the real world also presents situations of *continuous choice*, where there is an infinite number of strategies differing by inappreciable degrees. For example, if you are planning the budget of a Canadian election campaign, you could think of spending any amount between a minimum of zero and the legal maximum of about $12 million. With so many choices, it makes more sense to think of the utility function as continuous rather than as a set of discrete options.

Continuous functions are often represented by lines. Chapter 5 provided an illustration of using a line to represent a situation of continuous choice, in the discussion of Robert Axelrod's theory of the minimum connected winning coalition. There, the ideological and ethnolinguistic positions of different parties were represented by points on a line. In general, this kind of graphic representation is known as a *spatial model*, because a single line creates a one-dimensional space, a pair of lines intersecting at right angles creates a two-dimensional space, and so on.

The Median Voter Theorem

Spatial models were first used in economics to analyse locational decisions made by firms. Anthony Downs pioneered the application of spatial models to politics, using them to represent the 'locational' decisions

Invasion from the Right: Reform Party in 1993 Election 141

made by political parties when they take positions in order to compete against each other in elections. Downs's simplest model is also his best known; it predicts that two parties competing against each other will converge on the position of the median voter. The result is at least slightly paradoxical because it means that two competing parties will tend to become like Tweedledum and Tweedledee, offering only an illusionary choice to the voters.

The Downsian model of two-party competition is based on a number of assumptions, some of which are necessary to all spatial models:

- All voters have political positions that can be represented as points in a geometric space.
- Political parties assume pointlike positions in an attempt to appeal to voters.
- Voters, as rational actors, vote for the party located closest to their own ideal point. They use the party's location as a guide to the sort of policies it would enact if it became the government, and they naturally want policies as close as possible to their own preferences.

Of course, these assumptions will not apply fully in the real world. Some voters may have so little interest in politics that they do not have positions. Parties, because they are large and internally diverse organizations, may not be able to adopt distinct, pointlike positions; their 'positions' may be quite blurry and vague, more like a band or cloud of points. And voters may choose among parties for reasons other than policy, such as family tradition or respect for honesty and competence. But assumptions do not have to apply perfectly in order to be useful. If they apply fairly well to a considerable fraction of voters, they can generate a model with some degree of explanatory and predictive power.

Downs also used a second set of assumptions in his simplest model. These assumptions are clearly applicable only in some situations and could be varied to create a family of models of party competition:

- There is only one significant dimension of political difference in the society, namely the ideological spectrum of left and right. This statement may be true for some countries at some times; but politics is often multidimensional, involving conflicts over issues such as language, ethnicity, and religion that cut across ideological cleavages.
- There are only two parties competing for power. This axiom is hardly ever true in the strict sense, but it is often a reasonable

approximation of the truth for practical purposes. Ross Perot ran as a third-party candidate for president in the U.S. elections of 1992 and 1996, but only the Democratic and Republican candidates had any realistic chance of winning those contests.

- The distribution of voters' positions approximates a normal curve, with most being clustered toward the centre and far fewer lying toward the extremes. This description seems applicable to most societies at most times, but it is not universally true. Public opinion can become polarized into a bimodal distribution with only a few people in the centre and many more lying toward the extremes. But under such circumstances, democracy will probably break down into civil war, and spatial models of party competition will be irrelevant until order is restored.

Downs's simplest model is depicted in Figure 9.1.

The convergence of the two parties A and B on the position of the median voter can be informally explained as follows: Voters vote for

FIGURE 9.1
Simplest Model of Two-Party Competition

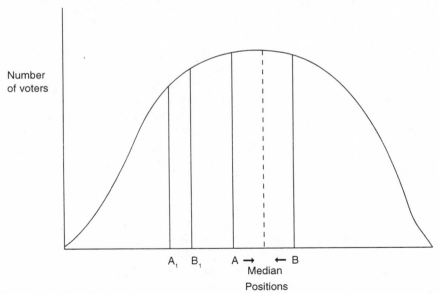

the party nearest their own position. If a party positioned itself to the left of the median, like party A_1 in Figure 9.1, the second party could position itself immediately to the right of the first, as B_1 has done. B_1 will now get the votes of everyone to the right (more than half the population), A_1 will get the votes of everyone to the left (less than half), and B_1 will defeat A_1 in the election. Hence, no party will ever want to position itself elsewhere than at the median, because to do so allows the other party to cut it off from a majority of voters.

The median voter theorem can also be expressed more formally as the minimax solution of a zero-sum game. Assume that A has three choices: to position itself to the left of the median, at the median, or to the right of the median. B also has three choices: to position itself marginally to the left of A, at the median, or marginally to the right of A. B would never think of going farther than marginally away from A in either direction because there is no advantage in doing so. If you get to one side of your opponent, no matter by how small a margin, you get all the voters lying in that direction because you are now closer to them than your opponent is.

Let 1 represent the payoff for winning (getting more votes than your opponent), −1 the payoff for losing (getting fewer votes), and 0 the payoff for a tie (getting the same number of votes). To see how the 3 × 3 matrix in Table 9.1 was filled in, consider the upper-left cell. In this scenario, A chooses to be left of the median, and B (foolishly) chooses to go to the left of A. Under these assumptions, A wins because B is cut off from more than half the voters, and the payoff vector is (1, −1). All payoffs are arrived at by a similar process of reasoning.

For each party, positioning itself at the median is a weakly dominant strategy, so the strategy pair (Median, Median) is the saddlepoint of the game.

TABLE 9.1
Normal Form of Two-Party Competition Game

		B		
		Left of A	At the Median	Right of A
A	Left of Median	1, −1	−1, 1	−1, 1
	At the Median	1, −1	**0, 0**	1, −1
	Right of Median	−1, 1	−1, 1	1, −1

Although Downs's first model is admittedly simplistic, it correctly predicts the competition between two large centrist parties that has generally characterized politics in the five 'Anglo-Saxon' democracies (Britain, Canada, New Zealand, Australia, and the United States). All these countries have had electoral systems that restrict competition to two large parties,[1] and those two parties have tended to converge upon more or less the same positions. To take a recent Canadian example, the Liberal opposition heavily criticized some of the main initiatives taken by Brian Mulroney's Conservative government, such as free trade with the United States, the Goods and Services Tax, a low target rate of inflation, and privatization of Crown corporations; but the Liberal government of Jean Chrétien did not depart from any of these policies and even pushed some of them further. In New Zealand, it was the Labour Party that began downsizing government after the 1984 election, but the National Party continued and intensified the policy after it came to power in 1990. To be sure, there have been periods of polarization, as between the British Conservatives and Labour when Margaret Thatcher was Tory leader; but such episodes have usually lasted only a few years and have been followed by a return to ideological convergence, as happened in Britain when both major parties changed their leaders. Even if the Downsian model is not entirely accurate, it is clearly a useful first approximation.

Also, it is not hard to build greater degrees of realism into the model. For example, the ideological position taken by a party is more like a line segment than a point; that is, it cannot be precise. And, in a world of imperfect information, having a broader position may allow one party to overlap with a competing one and thus attract some of its support. Moreover, party leaders do not have perfect freedom to position the party wherever they wish in order to attract votes; they also have to pay special attention to the views of the people who volunteer time and money to keep the party going. These volunteers and donors tend to be more extreme, in one direction or another, than the median voter; it is the very fact that they are off centre, so to speak, that creates an incentive for them to volunteer time and money to modify the 'natural' equilibrium at the position of the median voter. Volunteers and donors, therefore, tend to pull their parties away from the centre, and leaders must balance that pull against the imperatives of vote maximization.[2] Adding all this to the simplest model gives a more realistic one in which two parties compete with overlapping but discernibly different positions (Figure 9.2).

FIGURE 9.2
Refined Model of Two-Party Competition

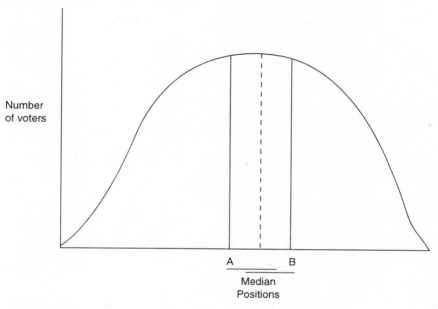

Third Parties[3]

If we relax the constraint that there can be only two political parties, is there a strategy by which a third party can enter the system and even displace one member of the duopoly? One possibility comes from a model developed by Steven J. Brams for American presidential primaries and extended by Réjean Landry to the case of party competition in a parliamentary system. Assume that a new party C is trying to break into a system dominated by old parties A and B. One obvious move would be for C to position itself just to the right of B (to the left of A would amount to the same thing).[4] If such a move were possible, C would be closer than B to most right-wing voters and should attract their support, thus finishing ahead of B. How many members C would elect would depend (in a first-past-the-post voting system like Canada's) upon the geographical concentration of the conservative voters for whom it was contesting with B. If such voters were evenly dispersed across many electoral districts, it is possible that C's challenge would do nothing but produce a landslide for A, whose left-wing support would be

FIGURE 9.3
Brams/Landry Model of Third-Party Entry

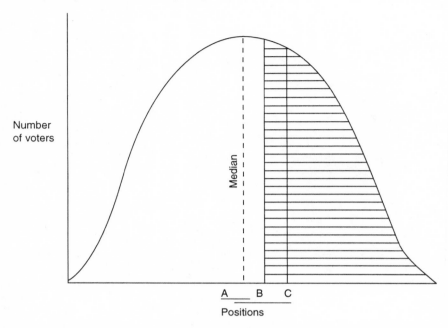

unaffected. But whether or not C can elect many members the first time, the model suggests that it should be able to outflank and finish ahead of B, thus positioning itself to enter the duopoly in the future (Figure 9.3).

I call this model 'Invasion from the Margin.' In its simplest form, it is, of course, patently unrealistic. It predicts wave after wave of successful invasions from both left and right, leading to a virtual kaleidoscope of parties. In fact, politics does not look like that anywhere in the world; the tendency is always for a small number of parties to assume long-term dominance.

In addition to plurality voting, there are two other factors blocking 'Invasion from the Margin': inertia and imperfect information. Like any purveyor of goods and services, an established political party has a huge advantage in reputation, credibility, and name-recognition over a new competitor. To have any hope of success, the recent entrant must differentiate itself from the established party. That task requires assuming a position not too close to the duopoly; for if C is only slightly different from B, why would voters who are used to B take the risk of supporting C, about which they know very little? But taking a position

Invasion from the Right: Reform Party in 1993 Election 147

far out toward the tail of the opinion distribution in order to promote clear differentiation carries its own risks. The end of the spectrum harbour extremists whose active support can be counterproductive for winning elections: communists and anarchists on the left, racists and fascists on the right. Thus, a new party playing 'Invasion from the Margin' must find a position far enough away from its main competitor to differentiate itself – but must also draw an effective line beyond itself so that it does not get discredited by extremists.

Finally, even if the new party finds a workable position, its main competitor can respond by moving toward it. Since the established party is by hypothesis operating in the region of the opinion distribution where voters are most numerous, it needs to take only a relatively small step away from the centre to win back a substantial number of voters who might be attracted to the new party. According to the logic of the spatial model, voters lying to the right of the midpoint of line segment BC in Figure 9.3 should vote for C, while voters lying to the left of the midpoint should vote for B. A move by B to the right also shifts the midpoint of BC to the right, thus transferring some voters from C back to B. For this reason, Downs, who was aware of the idea of 'Invasion from the Margin,' dismissed it as unworkable. In his opinion, it would not enable a new party to break into the system, though it might succeed in moving the position of an existing party away from the centre, at least for a time.[5]

Downs's rejection of 'Invasion from the Margin' may be valid for U.S. politics, where the barriers to new-party entry are extraordinarily high; but it is untenable as a general proposition. There is at least one modern example of its success in Canada – the CCF/NDP, which established itself in Canadian politics by outflanking the Liberals on the left. As the model predicts, it had problems with extremists on the far left, but it eventually managed to drive out most of the communists and fellow travellers. It vaulted over the barrier of first-past-the-post voting by relying upon the votes concentrated in the working-class neighbourhoods of major cities (Vancouver, Winnipeg, Toronto); manufacturing centres (Windsor, Hamilton, Oshawa); unionized natural-resource-extraction sites (mining, forest products); and wheat farmers in Saskatchewan.

The Reform Party has not yet exhibited the endurance of the CCF/NDP, but its spectacular entry into Canadian politics in the election of 1993, when it took 19 per cent of the popular vote and won 52 seats, requires analysis. In the balance of this chapter, we will attempt to test 'Invasion from the Margin' as an explanation of the Reform Party's 1993 success.

The Reform Party: 'Invasion from the Right'

The Progressive Conservatives in the 1980s espoused a number of positions shared by the Liberals and New Democrats: official bilingualism, multiculturalism, deficit spending, medicare and other social policies, and several waves of attempted and failed constitutional change (the Charter, Meech Lake, Charlottetown). This direction led to the virtual nonrepresentation of strongly conservative voters. It is not surprising, therefore, that the 1980s saw several attempts to found a federal party to the right of the Conservatives. Seen in this perspective, the foundation of the Reform Party in 1987 was the third act of a drama that began with the formation of the Confederation of Regions Party in 1983 and carried on with the establishment of Christian Heritage in 1986.

There is no doubt that Reform Party members see themselves as strongly conservative. A mail survey of delegates to the party's 1992 national assembly showed that they considered themselves well to the right of the Progressive Conservatives. On a seven-point scale ranging from 1 (extreme left) through 4 (centre) to 7 (extreme right), they scored themselves on average as 5.3 (strongly right of centre) and the PCs as 3.8 (slightly left of centre).[6] Eighty per cent labelled themselves conservative and 86 per cent saw the Reform Party as conservative.[7] About three-quarters of Reform members have never belonged to another federal party – but of those who have, 73 per cent used to belong to the PCs and 7 per cent to Social Credit.[8] These are the volunteers and donors without whom the party would quickly collapse.

There is, however, one problem in interpreting the Reform Party as mounting an 'Invasion from the Right': the party's leader, Preston Manning, does not see it that way. When pressed for an ideological designation, Manning sometimes calls himself a 'social conservative,' the term used in the book *Political Realignment*, which he helped his father write in 1967 when Ernest Manning was still premier of Alberta.[9] But more often, Manning denies that the terms left, right, and centre have any relevance to contemporary politics. He said as much at his speech at the Vancouver Assembly of May 1987, when he called on those in attendance to found an 'ideologically balanced' new party with 'a strong social conscience and program as well as a strong commitment to market principles and freedom of enterprise,' a party of 'hard heads and soft hearts, able to attract supporters away from the Liberals and NDP as well as the Conservatives.'[10] He repeated these sentiments in his 'Hockey Analogy,' which he mailed to all party members in January

1990. There he compared the party to a hockey line of three forwards playing right (free enterprise), centre (populism), and left (social concern). The crucial thing in his view was to integrate these ideological perspectives; for 'it is a virtual certainty that the politics of the 21st century will not be oriented on a right-left-centre basis.'[11] Manning continued to say similar things in public down through the 1993 election campaign.[12] However, in spite of the leader's statements, the Reform Party is considered, not only by its own members, but also by most political observers, to be on the right.

If 'Invasion from the Right' was the Reform Party's major operational strategy in the 1993 election, the following statements should be true:

1 On one or more issues that were central to the campaign, the Reform Party would position itself to the right of the major party (the Progressive Conservatives) that previously stood farthest to the right.
2 The Progressive Conservatives would have to meet the challenge by either (a) moving to the right to recapture defectors, or (b) moving to the centre to attract new voters from other parties. A choice of (a) should hurt Reform, while a choice of (b) should help it.
3 There would be evidence that Reform's success was due to its positioning on the right rather than to other factors.
4 Reform would tend to do well in electoral districts where the Conservatives did well in the past.
5 Given the ideological location of the other parties, Reform would attract more voters who previously voted Conservative than those who voted Liberal, and more who voted Liberal than who voted NDP.
6 Those who voted Reform would be more ideologically conservative than those who voted for other parties.

The balance of this chapter examines evidence from the campaign and the election relating to these six predictions. The evidence will show that all are strongly confirmed.

The 1993 Campaign

The Conservatives were in a virtual dead heat with the Liberals when the writ was dropped (34 to 33 per cent, according to Environics, 36 to 37 per cent according to Angus Reid), and far ahead of Reform.[13] However, as is now widely recognized, the 1993 Conservative election campaign was the most incompetent in Canadian history. Yet this collapse need not have benefited Reform; the alienated Tory voters could have gone en masse to the Liberals and the Bloc Québécois. The fact that Reform did benefit substantially is owing to the precise nature of one of Prime Minister Kim Campbell's major decisions – the way in which she vacated the right of the political spectrum, leaving Reform as the best option for 'small-c' conservative voters.

Initially, Campbell appeared to want to occupy the right by making the deficit her main theme.[14] But she came under increasing pressure as she refused to discuss the details of how she would fulfil her leadership campaign pledge to balance the budget in five years. Her undocumented position on this issue looked vague in comparison to Reform's 'Zero in Three' paper or even the Liberals' *Red Book*. This pressure led to a series of contradictory statements that demolished her credibility on fiscal responsibility, the main issue for conservative-minded voters in the election.

On 20 September Campbell said that, contrary to earlier statements, she would release some details of her deficit-cutting plans.[15] But only three days later, while confirming that her government intended to 'completely rethink our system of social security,' she refused to discuss the substance of the issue: 'You can't have a debate on such a key issue as the modernization of social programs in 47 days. ... [An election campaign] is the worst possible time to have that discussion ... because it takes more than 47 days to settle anything that is that serious.'[16] The next day, however, she partially backtracked by promising to set out 'the principles that I believe must guide any useful debate on how we as a country must modernize our social programs.'[17]

On 27 September, Campbell did put some deficit-reduction numbers on the table,[18] but it quickly became evident that they did not add up. In particular, she revealed in a visit with the editorial board of the *Globe and Mail* that she did not understand the difference between a decrease in the annual budget and a cumulative saving over five years.[19] Then, during the English-language leaders' debate, she was unwilling or unable to answer Lucien Bouchard's pointed question about how large the current deficit was estimated to be: 'A simple figure. What is the real deficit?'[20]

Immediately after the debate, the Tories launched a series of attack ads against the Reform Party, using the image of a magician sawing a woman in half to satirize Manning's 'Zero in Three' deficit-reduction program.²¹ In coordination with this campaign, Campbell began to attack Manning as 'a right-wing ideologue who has completely lost sight of ... the values that we have to preserve [in] our social programs and to create a caring society.'²² Emphasizing her new role as defender of the welfare state, she told Peter Gzowski, the host of the CBC's *Morningside* radio program, that she 'would throw [herself] across railway tracks to save the health care system.'²³

The net result was that, during a three-week period in the middle of the campaign, Campbell vacated the right of the political spectrum. Although she continued to criticize the Liberals for fiscal irresponsibility, her position lacked conviction because she was simultaneously posing as the defender of social programs against Reform. She thus marched away from the traditional ground that the Conservatives had always occupied.

The data show that the period when Campbell was vacating the right was precisely the time when Reform support shot upward. Figure 9.4 is a compilation of national polls conducted during 1993.²⁴ It shows that Reform support during the campaign was static at 10–11 per cent until about 20 September, then rose quickly to 17–18 per cent by the end of September and stayed at that level up to election day. Correspondingly, Conservative support held at about 35 per cent until 20 September, then fell off precipitously to the low 20 per cent range by early October. It also dropped sharply again in the last week, at the time of the Tory attack ads featuring Chrétien's facial distortion and of Campbell's public criticism of her former cabinet colleagues Brian Mulroney, Jean Charest, Don Mazankowski, and Robert de Cotret.²⁵

The timing of these developments is crucial. Although Campbell made mistakes from the very beginning, Reform did not move ahead until Campbell began to vacate the right. Similarly, the Conservative mistakes at the end of the campaign, which although serious were not ideological in character, moved voters to the Liberals and the Bloc Québécois, not to Reform. Reform could profit from Campbell's mistakes because, at the crucial time, Manning's statements had the effect of positioning the party on the right. He played 'Invasion from the Right' with perfect timing.

As planned, Manning began the campaign with a 'let the people speak' phase.²⁶ He pursued this theme for about a week in an initial tour of Canada's major cities. Although the party did not move ahead in the polls, it may have been a useful exercise, allowing him to reconnoitre the landscape, so to speak, while he crisscrossed the country.

152 Game Theory and Canadian Politics

FIGURE 9.4
1993 Poll Results

The Voters Speak
A look at how major political parties have fared in public opinion polls*

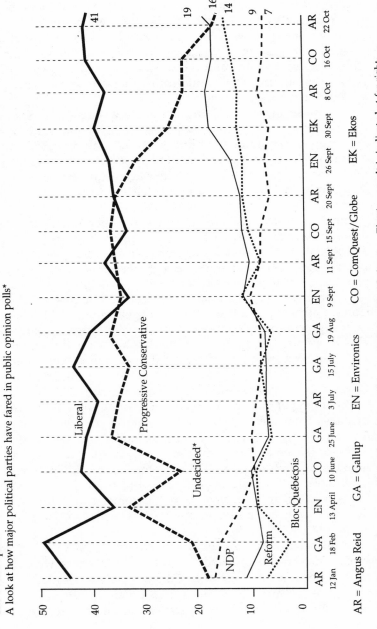

*Percentages for parties have been calculated without including undecided voters. Election result is indicated at far right.
Source: From Canadian Press and Globe and Mail

After about 10 days of campaigning, he veered sharply to the right. The turning point came when he re-released the 'Zero in Three' plan in a speech on 20 September in Peterborough, Ontario. Setting up an empty chair for Kim Campbell, Manning quipped: 'We have done all the homework on this and all she has to do is to take notes.'[27] The *Globe and Mail* again endorsed the plan, as it had done in the spring, giving Reform's campaign a major boost.[28]

The 'Zero in Three' package, released at the time that Campbell was giving up the ground of deficit reduction, was bound to position Reform firmly on the right. It was so vigorously attacked by the other parties that it took on a life of its own and set the tone for the rest of the campaign. In being forced to defend various aspects of 'Zero in Three,' particularly the controversial cuts to old age pensions and unemployment insurance, Manning had to appear as a conservative critic of social programs.

Of course, 'Invasion from the Right' never had the potential to win enough seats to form a government. Throughout the campaign, concern about the deficit was the top priority of only a minority of voters. In a ComQuest poll carried out 11–14 October, 57 per cent of respondents said that 'the government should invest money in job programs and training programs, even if it means increasing the deficit,' while 31 per cent said that 'the government should concentrate on reducing the deficit, even if it means more unemployment.'[29] Around the same time, Environics, using different wording, found that 41 per cent of respondents named unemployment as the most important issue, against 22 per cent who named the deficit.[30] In emphasizing the deficit, Reform was appealing only to a minority of voters – but a minority concentrated geographically in Alberta and British Columbia,[31] and demographically in small towns and middle- to upper-income suburbs. The concentration was sufficient to allow the party to win seats if it could come to 'own' the deficit issue, as Reform eventually did. The strategy worked well for breaking into the system, although a broader base of support is necessary if Reform is ever to fulfil Manning's dream of forming a government.

In the final phase of the campaign, Manning put much emphasis on the idea of minority government. He repeatedly called upon voters to deny the Liberals a majority, thereby letting Reform hold the balance of power and act as the 'fiscal conscience' of Parliament.[32] He also at times suggested that voters should make Reform the official opposition by giving them more seats than the Bloc Québécois: 'It is absolutely imperative that the balance of power in any minority Parliament be held by federalists rather than separatists. This is the way Reform can "beat the BQ" even though we are not yet present in Quebec.'[33] But even though Manning pushed these themes hard in the final days, they do not seem to have attracted any further support; Reform's vote share remained at the level achieved in the middle of the campaign, confirming that positioning on the right was the key to its success in the election.

This brief review of the campaign supports the first three predictions derived from the model of 'Invasion from the Right':

1 Reform positioned itself to the right of the Conservatives.
2 The Conservatives reacted by moving to the centre, which benefited Reform.
3 Reform's support rose when its ideological positioning was most visible to the public. Its support was static when Manning was stressing other issues not relevant to the left-right spectrum (populism at the beginning of the campaign, strategic voting at the end).

Who Voted Reform?

Of the 52 seats won by Reform candidates in 1993, 35 had been won in 1988 by Conservatives, and 17 by New Democrats. Superficially, it seems as if Reform did well on both PC and NDP territory, thus raising doubts about 'Invasion from the Right.' However, survey data show that, while the Conservative connection is valid, Reform actually attracted very few crossover voters from the NDP. Reform was able to take seats from the NDP because the NDP's support collapsed, but that does not mean that previous NDP voters themselves went over to Reform.

That Reform did well on traditional Conservative territory is corroborated by a closer look at Ontario, where Reform won only one seat and did not come within 4000 votes of winning any others. Reformers did finish second in 56 Ontario ridings, but many of these second-place showings were so far back as to be meaningless. A more significant indication of Reform strength in a riding was to get at least half as many votes as the winner. This happened in 23 Ontario constituencies, of which 22 had been Conservative in 1988 and one had been NDP – Ed Broadbent's old riding of Oshawa.

Table 9.2 displays the correlation coefficients between the 1993 Reform percentage of the vote in each electoral district and the percent-

TABLE 9.2
Correlation of 1993 Reform Vote Percentages by
Constituency with 1988 Percentages of Various Parties

Province	Liberal	NDP	PC	Reform	(PC + Reform)
BC	−.45*	−.20	.45*	.47*	.61**
Alberta	−.71**	−.79**	.52*	.63**	.86**
Sask./Man.	−.50*	−.20	.87**	.54*	.84**
Ontario	−.60**	−.14	.61**	–	.61**
All	−.79**	−.07	.60**	.73**	.80**

* p < .01
** p < .001

ages of the votes obtained by the various parties in the same ridings in 1988.[34] I have also added the 1988 Reform and PC votes together to create another variable in the 72 ridings in which there was a Reform candidate in 1988. The correlations are calculated separately for each province and then collectively for all provinces together. Because of their small size, Saskatchewan and Manitoba are treated as if they constituted a single province. Atlantic Canada is not included in the analysis because Reform contested only 20 of 32 seats in the region and nowhere finished higher than third place; its support in this region was too tentative for much profit to be derived from detailed analysis. Quebec is not included because it did not present any Reform candidates.

The most striking finding is that in each province, as well as in all provinces taken together, there was a strong positive correlation between the 1993 Reform vote and the 1988 PC vote. This correlation confirms that Reform did well in ridings where the Conservatives used to be strong. It is also noteworthy that the correlation can be increased to 0.80 by treating the 1988 PC + Reform votes as a single variable (that is, as two wings of an already dividing bloc of conservative-minded voters). In simple terms, Reform did well where the PCs used to do well because Reform appealed to the same kinds of voters.

Not surprisingly, there is an almost equally strong negative correlation (−0.79) between the 1993 Reform vote and the 1988 Liberal vote. Again, in simple language, this finding means that Reform had trouble attracting support in Liberal territory; the kinds of people who have historically voted Liberal (Roman Catholics, Jews, francophones, visible minorities, urban 'sophisticates') obviously did not respond well to Reform's appeal. Interestingly, however, the correlation between the

1993 Reform vote and the 1988 NDP vote is strongly negative only in Alberta (–0.79) and is virtually zero overall (–0.07). Although this finding requires further investigation, it probably reflects the fact that there are several distinct types of NDP voters – public sector workers, unionized industrial and resource-extraction workers, prairie grain farmers, the urban poor, ideological activists (feminists, environmentalists, gay rights advocates, etc.) – distributed in a highly uneven way across the country.

Because of the well-known ecological fallacy, it is not safe to use data about *aggregates* – in this case, electoral districts – to make inferences about the behaviour of *individuals*. Fortunately, we can supplement the riding data with data about individuals drawn from a national sample survey ($n = 1496$) conducted by Harold Clarke immediately after the election.[35] Clarke asked respondents how they voted in 1988, which brings us to a test of the fourth prediction of 'Invasion from the Right.' Did Reform voters come more from former Conservatives than from former Liberals, and more from former Liberals than from former NDP supporters? The answer is contained is Table 9.3.

Outside Quebec, where Reform did not present any candidates, 38 per cent of those who voted Conservative in 1988 voted Reform in 1993, as compared to a 15 per cent defection rate from the Liberals and 11 per cent from the NDP.[36] In general terms, Reform clearly drew more from the right than from elsewhere. On the other hand, recruitment from the Liberals and NDP was more than negligible; votes from these two parties were essential to winning some close races. In that sense, Manning's depiction of Reform as more than a party of the right may have had some payoff if it increased the rate of defection from the Liberals and the NDP.

TABLE 9.3
1993 Vote Percentage by 1988 Vote Percentage (Omitting Quebec)

		1988 Vote			
		NDP	Liberal	PC	Reform
1993 Vote	NDP	43	3	3	0
	Liberal	41	79	26	0
	PC	5	3	33	0
	Reform	11	15	38	100
n =		202	228	323	11

TABLE 9.4
1993 Vote Percentage by 1988 Vote Percentage (Four Western Provinces)

		1988 Vote			
		NDP	Liberal	PC	Reform
1993 Vote	NDP	61	2	6	0
	Liberal	26	66	18	0
	PC	4	4	23	0
	Reform	9	28	55	100
n =		64	53	162	11

It is also worth looking in Table 9.4 at retention and defection rates just in the four western provinces, where Reform won 51 of its 52 seats. In the West, the Conservatives lost a massive 55 per cent of their vote to Reform, and the Liberals 28 per cent, as compared to only 9 per cent for the NDP.[37] Again, the story is mainly one of recruitment from the Conservatives, but with a useful supplement from the Liberals and to a lesser extent the NDP.

Nationally, fully 70 per cent of those who voted Reform in 1993 had voted either Conservative or Reform in 1988. (In this survey, all respondents who had voted Reform in 1988 repeated in 1993, but the number was very small, only 11). For the West, the percentage of 1993 Reform voters who had previously voted Conservative or Reform rises to 81 per cent. The Reform vote, especially in areas where the party was successful, was basically a secession movement from the Conservative Party, starting in a small way in 1988 and reaching large proportions in 1993. Switchers from other parties were a useful supplement in tight races, but were not numerous enough to affect the character of the coalition.

These data also show that the alleged shift from the NDP to Reform in British Columbia was illusory. Only 9 per cent of those in western Canada who voted NDP in 1988 changed to Reform in 1993; and in British Columbia, according to Clarke's survey, the figure was even smaller, only 8 per cent. Almost three times as many defecting NDP voters went to the Liberals as to Reform. True, the NDP lost 15 seats to Reform, but not through direct vote transfers. While the NDP was losing votes mainly to the Liberals in British Columbia, Reform was holding the vote it had achieved in 1988, picking up more than half the 1988 Conservative vote and almost a third of the 1988 Liberal vote, but adding less than a tenth of the 1988 NDP vote.

TABLE 9-5
Support for Selected Issues by Vote, 1993

	per cent	
	PC	Reform
Minorities: More should be done for racial minorities	29.5	16.8
Morality: Women should stay home with children	20.5	29.6
Welfare state: Welfare state makes people lazy	28.9	47.1
Deficit and spending cuts: Support programs over cutting the deficit	22.7	16.4

Overall, Clarke's data confirm the fourth and fifth predictions, namely that Reform did well where the Conservatives used to do well, and Reform drew its electoral support more from the Conservatives than the Liberals, and more from the Liberals than the NDP.

Survey data from another source confirm the sixth prediction of 'Invasion from the Right,' namely, that Reform voters would be more ideologically conservative than voters of any other major party, and in particular more conservative than those who voted Progressive Conservative. The evidence comes from the 1993 Canadian Election Study conducted by Richard Johnston and associates.[38] Table 9.5 shows the percentage of voters in this study who responded to the particular items in what would generally be considered the conservative direction.[39] On each of the four items reported here, Reform voters were clearly more conservative than Progressive Conservative voters. Moreover, these items were drawn from a larger set of 13, and Reformers were to the right of the PCs on 12 of the 13.[40] The evidence is overwhelming that in 1993 Reform drew its support from the most strongly conservative of Canadian voters.

Other Models of Entry

'Invasion from the Right' is not the only strategy with which the Reform Party has experimented. It has also made several attempts to establish dimensions of political competition other than the standard left-right ideological dimension. In Downsian terms, this initiative means establishing a second dimension of cleavage, thus creating a two-dimensional political space with more room to manoeuvre. Although ideological conflict is ubiquitous in democratic politics, there is in principle an unlimited number of potential issue dimensions.[41] Riker argues that political entrepreneurs on the margins are constantly trying to raise new issues, seeking dimensions of cleavage that will pry apart existing coalitions. He sees a 'natural selection' of issues in which most such attempts fail but an occasional one succeeds in bringing about a major realignment.[42]

Beyond 'Invasion from the Right,' Reform has experimented with at least four other models of new-party entry, each of which involves establishing a second dimension of conflict cutting across the purely ideological one.

'The Party of the West'

When first established, the Reform Party took as its motto 'The West Wants In' and ran candidates only in the four western provinces in the 1988 election. At this stage, it strongly emphasized regional issues, such as the Triple-E Senate. It decided to go national in 1991, and except in one or two speeches, Manning did not mention regional issues in the 1993 election. But even though it is no longer 'The Party of the West' in a formal sense, it still carries something of that identity.

This association was obviously an important factor in the 1993 election. Of Reform's 52 seats, 51 were won in the West: 24 in British Columbia, 22 in Alberta, 4 in Saskatchewan, and 1 in Manitoba. The result was partly, but only partly, because the Conservatives had been strong in the West. Clarke's survey also shows that a greater proportion of Conservatives came over to Reform in the West than in Ontario or Atlantic Canada (Atlantic Canada, 8 per cent; Ontario, 26 per cent; Prairies, 54 per cent; British Columbia, 56 per cent). Beginning life as a regional party greatly helped the Reform Party to achieve the territorial concentration necessary to win seats in the Canadian first-past-the-post electoral system. Ironically, however, Reform did not succeed in winning western seats until it had become a national party. One can only wonder whether Reform could have won as many western seats in 1993 if it had remained a western party.

'The Party of the Hinterland'

Preston Manning's original strategy was to expand the Reform Party into a national party by attracting support in the 'resource-producing regions' of the 'hinterland' – 'not only in western Canada, but in Atlantic Canada and in the rural and northern areas of Quebec and Ontario.'[43] However, there is no sign that this strategy explains Reform's success in the 1993 election. Reform did poorly in Atlantic Canada, northern Ontario, the Northwest Territories, and Yukon. Although Reform has strong rural support in Ontario and the West, that support is not located in the remote resource-producing 'hinterland'; it is more of a small-town and agricultural phenomenon. And Reform also did very well in the affluent outer suburbs of metropolitan Toronto and Vancouver.

'The Party of the People'

Manning's favourite notion is that the Reform Party is a 'populist party' inheriting the tradition of earlier Western populist parties.[44] This belief ties in with his rejection of ideology, for there have been earlier populist movements of the left (CCF/NDP), right (Social Credit), and centre (Progressives). A populist issue dimension amounts to emphasizing conflicts between the 'elite' and what Manning likes to call 'the common sense of the common people.'[45]

Although populism is the most important thing in Manning's mind, and it undoubtedly motivates many Reform activists, there is no evidence that it is the prime determinant of Reform's electoral support. If Reform were really 'The Party of the People' as Manning claims, it would recruit members and supporters more or less evenly from the other parties. Yet all the evidence highlights the importance of defections from the Conservatives.

Be that as it may, populism continues to guide Manning's thinking. Shortly after the 1993 election, he drew up an expansion plan that called upon Reformers to recruit new support from the ranks of the NDP and the Liberals as well as the Conservatives, and to make the party 'reflect the demographics of the Canadian population as a whole' by attracting new support in precisely those categories where Reform has been weakest: women, young voters, visible minorities, and francophones.[46]

'The Party of English Canada'

The constitutional claims of Quebec played hardly any role in the founding of the Reform Party but were quickly brought to the fore as the party became known for its opposition to the Meech Lake Accord. At Reform's Edmonton Assembly in November 1989, Manning deliberately took a polarizing stand toward Quebec: 'If we continue to make unacceptable constitutional, economic, and linguistic concessions to Quebec at the expense of the rest of Canada, it is those concessions themselves which will tear the country apart .... A house divided against itself cannot stand.' Following Manning's lead, the assembly voted to adopt a hard-line position toward constitutional demands from Quebec:

The Reform Party supports the position that Confederation should be maintained, but that it can only be maintained by a clear commitment to Canada as one nation, in which the demands and aspirations of all regions are entitled to equal status in constitutional negotiations and political debate, and in which freedom of expression is fully accepted as the basis for language policy across the country. Should these principles of Confederation be rejected, Quebec and the rest of Canada should consider whether there exists a better political arrangement which will enrich our friendship, respect our common defence requirements, and ensure a free interchange of commerce and people, by mutual consent and for our mutual benefit.[47]

Although this strategy undoubtedly contributed to the rapid growth of the party between elections, it did not play much of an overt role in the 1993 election. Manning referred to it only obliquely, by asking voters to make Reform, rather than the Bloc Québécois, the official opposition. If 'The Party of English Canada' had been in play, Reform support should have been generalized ideologically, geographically, and demographically across English Canada, not concentrated as it was.

However, 'the Party of English Canada' has the potential to play a powerful role in a future election if relations between Quebec and the rest of Canada become a prominent issue. According to the 1993 Canadian Election Study, Reform voters are measurably less sympathetic to Quebec than are the supporters of any other party.[48] The results of that study are summarized in Table 9.6.

The 1993 election was really two elections held simultaneously. In Quebec, the main issue was sovereignty, and the electorate was polar-

TABLE 9.6
Attitude toward Quebec by Vote, 1993

	NDP	Liberal	BQ	PC	Reform
More should be done for Quebec (% agree)	21.1	27.1	83.4	21.3	6.8
Feelings for Quebec*	59.9	63.5	86.4	62.3	52.3

*average thermometer score: 0 is cold, 100 is hot, 50 is neutral

ized between separatists, who voted BQ, and federalists, who voted Liberal. Elsewhere in Canada, the main issue was the state of the economy (jobs or the deficit, depending on where you stood). Those worried mainly about jobs tended to vote Liberal, while those most worried about the deficit tended to vote Reform. But if a future election is ever held against the backdrop of a secession crisis, the polarization over sovereignty that now prevails in Quebec could very easily spill over into the rest of the country. Under such conditions, Reform might well strip away from other parties the voters most willing to engage in confrontation with Quebec.

In terms of spatial models, English-French relations constitute a powerful second dimension in Canadian politics that was latent in the 1993 election outside Quebec, but that cannot be neglected in a complete analysis. Impressionistically, the five parties could be located in this two-dimensional space as shown in Figure 9.5. I have located the Progressive Conservatives toward the Pro-French end because their current leader (at the time of writing) advocates distinct-society status for Quebec and rejects formulating a contingency plan – the so-called Plan B – to prepare for a possible separation crisis.

The overall conclusion is that 'the Party of the Right,' working with the residue of 'the Party of the West,' accounts for the Reform Party's success in the 1993 election. 'The Party of the Hinterland' is now only of historical interest. However, 'the Party of the People' and 'the Party of English Canada' are still of potential importance – the former because it dominates the mind of the party's leader, and the latter because Canada may still face a secession crisis in which politics would be polarized along linguistic lines.

Thus, the identity of the Reform Party is not fully fixed. Not long ago it was 'the Party of the West'; it was effectively 'the Party of the Right' in the 1993 election; it is 'the Party of the People' in Manning's mind;

FIGURE 9.5
Two-Dimensional Canadian Political Space

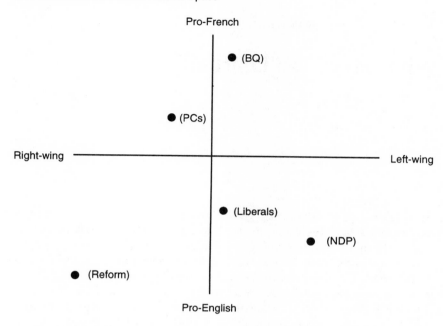

and it could easily become 'the Party of English Canada' in a secession crisis. Interestingly, this multivalence is an intrinsic part of Manning's own thinking on strategy, which to him is more a matter of timing than positioning. He has repeatedly characterized his strategy as 'waiting for the wave.' In his words, we 'keep positioning ourselves so that when the next wave comes along, we can ride it higher and longer.'[49] In this perspective, Manning's use of various strategies may represent attempts to ride particular waves as they come along.

10

What Have We Learned?

As a branch of mathematics, game theory is an intellectual edifice of assumptions and theorems, to be judged by standards of deductive logic. Like all of mathematics, game theory is a tautology whose conclusions are true because they are contained in the premises. Even if game theory told us nothing about the real world of government and politics, it would have its own beauty and integrity as an intellectual structure.

In this book, however, I have proceeded on the assumption that game-theoretic models can represent the real world of Canadian politics in an enlightening way. Game theory depicts rational actors seeking to maximize their own self-interest, and politics is also driven to a considerable degree by the pursuit of self-interest. Granted, political actors are altruistic on occasion, and the pursuit of self-interest is restrained by a combination of legal, moral, and customary norms. Nonetheless, the evidence of self-interest is all around us: interest groups seeking benefits for their members, political parties trying to win elections, voters responding to campaign promises, public servants seeking bigger budgets and career advancement. Even if self-interest does not make up the whole fabric of politics, it is at least the warp or the woof. Game theory, therefore, has the potential to highlight the threads of self-interest in the fabric of politics and the designs in which they are woven.

Theoretical plausibility, however, is not the same as empirical demonstration. That game theory ought to be useful does not mean it will be. Its explanatory value can be determined only by testing its models against real-world evidence. The time has come to review the applications made in the earlier chapters, to see what they contribute to an understanding of Canadian politics. Let us go over the results of the case studies one by one, divided into two groups: first, those that tease out the logic of institutional rules; second, those that model political behaviour.

Institutional Rules

Constitutional Amendment

The Banzhaf Power Index (BPI) is a measurement device, not an explanatory or predictive model. By computing the percentage of times that a province can be pivotal in approving or defeating a proposal, the BPI furnishes an objective index of voting power under a specific decision-making rule. However, because it lacks specific context, the BPI may miss the ways in which some provinces may have special power with respect to certain issues. For instance, it seems unlikely that any coalition, even one meeting the 7/50 requirements, would pass a distinct-society amendment against the opposition of Quebec.

Use of the BPI to measure provincial power under various amendment rules is a refinement of common sense. Informed observers understood the general drift of the five-region veto before anyone did the mathematical calculations. But the same could be said about any exercise in measurement. Looking at two people, I can tell that one is taller than another; but if I need to know exactly how tall they are, I pull out a tape measure.

As a measuring device, the BPI can help in the normative task of judging the fairness of decision-making rules. Judgment will depend on many factors, not just on the BPI; but having a precise *a priori* measure of the distribution of power should be helpful, even if not decisive. Rational choice and game theory thus help to answer institutional questions (what are the mechanics of the rules?) as well as normative questions (do we approve of these rules?).

The BPI could also be used in behavioural research. One could carry out BPI analysis of all the constitutional amendment rules discussed in Canada over the last 40 years, then test for correlations between provincial power shares under a particular rule and the degree to which provincial representatives advocated or opposed that rule. Research of this type might shed some light on the extent to which provincial politicians act on their perceptions of provincial interest in constitutional negotiations.

Leadership Selection

Rational-choice analysis of the run-off selection rule demonstrates a theoretical possibility that it may choose someone other than a Condorcet winner, which could be dangerous for the unity of the party. However, close study of three nominating conventions where this might have occurred did not turn up an actual example. In one case where it seemed particularly possible (the New Democrats in 1995), delegates who saw the possibility forestalled it by strategic voting. The defeat of a Condorcet winner remains a possibility, but the empirical evidence suggests it does not often happen in practice.

This application of rational choice complements existing institutional and philosophical approaches to the study of leadership selection. It provides important evidence to those who might wish to replace the run-off rule. It is also linked to the behavioural research now often carried out by those who administer survey questionnaires to convention delegates; it strongly suggests that the research instruments should contain items about preference order and strategic voting, as some have done in the past.

Parliamentary Procedure

Abstract analysis of the legislative process shows that the Condorcet extension favours the status quo. Against the backdrop of this knowledge, the Mulroney government's futile attempts to pass a new abortion law bring together institutional and behavioural approaches to the study of Parliament. They show how the government tried to manipulate the rules to get its favoured result and how backbenchers were able to fight back through strategic voting.

Of course, some elements of the story were always visible without the apparatus of game theory. All previous observers had recognized how the government had been thwarted by paradoxical coalitions of pro-choice and pro-life advocates. But the previous literature had not adequately explained the government's procedural manoeuvres before voting began, nor had it drawn sufficient attention to the Chicken game played between Mulroney and his backbenchers on the second try. Understanding legislative procedures in the light of rational choice leads to a more nuanced account of what happened, revealing a rich panorama of strategic behaviour normally suppressed by party discipline and bloc voting.

What Have We Learned? 167

There is also a normative dimension here with respect to that endlessly debated aspect of Canadian politics, party discipline. This detailed case study of how Parliament works in a rare example of free voting can help inform one's judgment of whether party discipline should be relaxed or abandoned. Although I personally still think that party discipline is exaggerated in Canada, this case study gave me a deeper appreciation of why party discipline is fundamentally desirable in a parliamentary system. I prefer decisions to arise from the will of the prime minister, who can be held politically accountable, than to emerge as unintended consequences of procedural rules, heresthetic manoeuvres, and strategic voting.

Behavioural Models

Lubicon Stalemate

Game theory in this instance provides an abstract model – Deadlock – to fit the behaviour of the main actors. What does this model contribute to our understanding? For one thing, it emphasizes the rationality of both the Lubicon and federal negotiators. Both sides emerge as tough, capable strategists pursuing hard-headed goals of great importance to their respective constituencies. This finding alone is a worthwhile contribution to the Lubicon debate, which has been carried out exclusively in moralistic terms. The friends of the Lubicon denounce the government as genocidal racists, while the friends of the government denounce the Lubicon as romantic obstructionists, out of touch with the modern world. In such an overheated climate, recognition of the essential rationality of one's opponents is a small step toward an eventual settlement.

The game-theoretical approach also exhibits some ability to predict the behaviour of the protagonists. If the Lubicon were purely outraged moralists, would they not take all avenues, including litigation, to pursue their goals? But in fact they avoid litigation. And if the government were really genocidal racists, they would not settle with any of the isolated communities. But in fact the government has made agreements with several neighbours of the Lubicon. The hypothesis of strategic behaviour seems to account for the facts better than the hypothesis of oppression and victimization.

It must be emphasized, however, that this is only a loose explanation *post hoc.* Already familiar with the historical facts, I selected Deadlock as the model to capture the salient features of the conflict, then did a little reasoning to see if behaviour on both sides was consistent with that model. However, Deadlock or any other model does nothing but formalize the preference orderings of the contestants. If one or both parties change their preferences, the game could turn into Chicken, Prisoner's Dilemma, or another of the dozens of 2 x 2 games identified in the literature.[1] And since the analyst has no independent knowledge of the actors' preferences, but has to infer them from words and actions, a change in behaviour means a change in model. When the parties eventually settle, we will need to find a new model to explain how the deadlock was broken.

This use of game theory models is far from the hypothetico-deductive paradigm of empirical science advocated by many exponents of rational choice. According to that paradigm, one would create models in advance by deducing how self-interested human beings would behave in particular situations, then test them against subsequently collected data. That may be the ideal; but, as Green and Shapiro have shown in *Pathologies of Rational Choice Theory*, rational choice has hardly ever lived up to the ideal.[2] I personally doubt that it is even a sensible ideal for political science.

Be that as it may, the analysis of the Lubicon conflict does not purport to be a test of a general law according to the hypothetico-deductive paradigm. The Deadlock model throws the spotlight on the role of self-interest in the conflict and shows how two rational, self-interested parties could prefer a stalemate to any available agreement. It does not uncover a general law of behaviour, but it makes this particular situation more intelligible by revealing the logic of both parties' actions. It is a modest, but still real, contribution to understanding a thorny conflict.

Metrication

The analysis of metrication is similar in principle to that of the Lubicon conflict – an *ad hoc*, retrospective exercise of finding a representative model, in this case, an n-person Assurance game. The empiricism is casual, consisting of personal observations of the degree of metrication in various domains of daily life. There is no actual measurement of the metric or imperial advantage in these domains and no actual data about people's perceptions of costs and benefits. I looked for such research but could not find it; metrication seems to be an overlooked, understudied topic.

Relying merely on casual empiricism, one can only say that the n-person Assurance model seems consistent with the little we know about metrication in Canada. But even that is worth saying, because it helps to clarify the logical structure of the metrication problem. Political philosophers have long pondered the question of what government should do in a free society. Attempting to characterize the role of government, John Stuart Mill wrote in his *Principles of Political Economy:* 'There are matters in which the interference of law is required, not to overrule the judgment of individuals respecting their own interest, but to give effect to that judgment: they being unable to give effect to it except by concert, which concert again cannot be effectual unless it receives validity and sanction from the law.'[3]

There is now a large game-theoretical literature on the nature of various collective-action problems. Determining whether the microfoundations of the problem approximate Coordination, Assurance, Chicken, or Prisoner's Dilemma leads to very different recommendations for the role of government.[4] Applying game theory to these problems lends logical precision to a traditional concern of political philosophy.

The Size of Coalitions

This analysis leads to two major conclusions: that victorious parties win by a greater margin under multiparty than under two-party competition, and that large winning coalitions are vulnerable, particularly when they are extended across a multidimensional issue space. The first conclusion is, as far as I know, new; one would never think to test that hypothesis without knowing Riker's size principle. Although the second conclusion has often been expressed by writers dissecting the fate of particular governments, game theory puts it in a systematic context.

Although both conclusions are supported by a reasonable amount of data, there is admittedly an *ad hoc,* circular character to the analysis because I was already familiar with the facts of federal political history before I thought of applying coalition theory to the interpretation. The conclusions would certainly be strengthened if they could survive a test against further bodies of data. One possibility for the first hypothesis would be to use provincial political history. A large body of data exists, since all the provinces from Quebec westward have undergone repeated episodes of both two-party and multiparty competition.

Provincial data could also provide a test for the second hypothesis, about the fragility of large coalitions. There have been many sweeping victories in provincial elections, including, for example, the 1987 New Brunswick election, when Frank McKenna's Liberals won 60 per cent of the popular vote and all of the 58 seats in the legislature. Interestingly, McKenna's support fell very little in the next two elections. The long run of overwhelming Social Credit victories in Alberta from 1935 to 1967 also comes to mind. A refined hypothesis worth exploring would be that large coalitions may be less fragile in provincial politics than in federal politics. Provincial coalitions may be less subject to cross-pressure in multidimensional issue space because each province is less internally diverse than the federation.

Another limitation of this analysis is that it does not attempt to test alternative explanations. There might be reasons other than coalition theory why the winning party's margin of victory has been consistently larger since Canada made the transition to multiparty politics. Some other model of multiparty competition could perhaps predict the same result. As with the principle of differential diagnosis in medicine, a theory is not securely established merely because it makes correct predictions; it must also make better predictions than other theories.

Entry from the Right

Of the various applications of game theory to Reform Party strategy in 1993, this one has the highest level of empirical support: it survives half a dozen different tests based on deductions from the model. To be sure, it does not give us a total explanation of the Reform Party's success in that election; but it highlights the factor of party positioning, which all analysts (and politicians) agree is important. This model furnishes a systematic way to think about positioning and strategy that could be applied to any election.

The 1993 election, however, was something of a special case because it was so one-dimensional, at least outside Quebec. The model of 'Entry from the Right' would not apply as effectively to the 1997 election, when Reform emphasized the French-English dimension of Canadian politics by running its famous 'Quebec-based politician' ads. In strategic terms, Reform tried to be not only the 'Party of the Right' in 1997, but also the 'Party of English Canada.' A two-dimensional model would be required to represent the major features of that election.

What Have We Learned? 171

As in the case of the Lubicon conflict, game theory offers a family of models, one of which fits the current facts; and when the facts change, you can pull another model off the shelf. What is missing, however, is a theory about when one or another model will be appropriate. Once again, game-theoretic models seem to be an elegant way of representing the key features of a particular situation, but they do not provide a full-fledged scientific theory allowing the analyst to predict when and why those facts will metamorphose into another configuration.

This limitation of game theory is also a limitation of all other theories in political science. There simply is no universally valid, highly predictive theory applying to all situations. I do not expect rational choice to become such a theory. I value it for the clarity with which it spotlights the effects of self-interest in many different settings, but I do not defend it as a 'theory of everything.' I prize every candle in the darkness of the universe, even if it is not a supernova of blinding illumination.

Notes

Preface

1 John von Neumann and Oskar Morgenstern, *Theory of Games and Economic Behavior* (Princeton, NJ: Princeton University Press, 1944).
2 Sylvia Nasar, 'Lost Years of a Nobel Laureate,' *Globe and Mail*, 19 November 1994.
3 Lloyd S. Shapley and Martin Shubik, 'A Method of Evaluating the Distribution of Power in a Committee System,' *American Political Science Review* 48 (1954), 787–92.
4 William H. Riker, 'The Entry of Game Theory into Political Science,' in E. Roy Weintraub, ed., *Toward a History of Game Theory* (Durham, NC: Duke University Press, 1992), 207.
5 My count; another observer might classify slightly differently.
6 I have made particular use of James D. Morrow, *Game Theory for Political Scientists* (Princeton: Princeton University Press, 1994); Avinash K. Dixit and Barry J. Nalebuff, *Thinking Strategically: The Competitive Edge in Business, Politics, and Everyday Life* (New York: W.W. Norton, 1991); Iain McLean, *Public Choice: An Introduction* (Oxford: Basil Blackwell, 1987); Steven J. Brams, *Rational Politics: Decisions, Games, and Strategy* (Boston: Harcourt Brace Jovanovich, 1985); Morton D. Davis, *Game Theory: A Nontechnical Introduction*, rev. ed. (New York: Basic Books, 1983); Henry Hamburger, *Games as Models of Social Phenomena* (New York: W.H. Freeman, 1979); and Anatol Rapoport, *Two-Person Game Theory: The Essential Ideas* (Ann Arbor: University of Michigan Press, 1966). Another recent treatment is Shaun P. Hargreaves Heap and Yanis Varoufakis, *Game Theory: A Critical Introduction* (London and New York: Routledge, 1995).

Chapter 1: Rational Choice

1 John Stuart Mill, 'On the Definition of Political Economy,' in *Essays on Some Unsettled Questions of Political Economy*, 2nd ed. (London: Longmans, Green, Reader, and Dyer, 1874; Augustus M. Kelley reprint 1968), 137.
2 Ibid., 139.
3 This is the position that Donald P. Green and Ian Shapiro call 'partial universalism.' *Pathologies of Rational Choice Theory: A Critique of Applications in Political Science* (New Haven: Yale University Press), 26–7.
4 Anthony Downs, *An Economic Theory of Democracy* (New York: Harper and Row, 1957), 260–76.
5 Mancur Olson, *The Logic of Collective Action: Public Goods and the Theory of Groups* (Cambridge, MA: Harvard University Press, 1965), 167.
6 Kenneth A. Shepsle and Mark S. Bonchek, *Analyzing Politics: Rationality, Behavior, and Institutions* (New York: W.W. Norton, 1997).
7 There are some further refinements. One can also say $a \geq b$, meaning that the chooser is at least indifferent between a and b, and may actually prefer a, but definitely does not prefer b to a. And similarly, one can say $a \leq b$. But we don't need to worry about these refinements now.
8 Green and Shapiro, *Pathologies of Rational Choice Theory*, 17–19.
9 Amos Tversky and Daniel Kahneman, 'The Framing of Decisions and the Psychology of Choice,' *Science* 211 (1981), 453–8; idem, 'Rational Choice and the Framing of Decisions,' in Karen Schweers Cook and Margaret Levi, eds., *The Limits of Rationality* (Chicago: University of Chicago Press, 1990), 60–89.
10 Jane J. Mansbridge, ed., *Beyond Self-Interest* (Chicago: University of Chicago Press, 1990).
11 Downs, *Economic Theory of Democracy*, 29.
12 Ibid.
13 Geoffrey Brennan and James M. Buchanan, *The Reason of Rules: Constitutional Political Economy* (Cambridge: Cambridge University Press, 1985), 36.
14 Howard Margolis, 'Dual Utilities and Rational Choice,' in Mansbridge, *Beyond Self-Interest*, 239–53; Mansbridge, 'Expanding the Range of Formal Modeling,' in ibid., 254–63.
15 Bill James, *Major League Handbook, 1996* (Skokie, IL: STATS Publishing, 1995), 303.
16 Bryan D. Jones, *Reconceiving Decision-Making in Democratic Politics: Attention, Choice, and Public Policy* (Chicago: University of Chicago Press, 1994), 68.

17 Charles Taylor, 'Impediments to a Canadian Future,' in *Reconciling the Solitudes: Essays on Canadian Federalism and Nationalism* (Montreal and Kingston: McGill-Queen's University Press, 1993), 187–201.
18 C.E.S. Franks, *The Parliament of Canada* (Toronto: University of Toronto Press, 1987).
19 Harold D. Clarke, Jane Jenson, Lawrence LeDuc, and Jon H. Pammett, *Absent Mandate: Canadian Electoral Politics in an Era of Restructuring*, 3rd ed. (Vancouver: Gage, 1996), 185.
20 Auguste Comte, *Cours de philosophie positive* (Paris, 1839), vol. 4, 200.
21 'Positive politics' *(politique positive)* and 'social science' *(science sociale)* were actually his first terms; they appear in unpublished writings as early as 1819. Auguste Comte, *Ecrits de jeunesse, 1816–1828* (Paris: Mouton, 1970), 467–82.

Chapter 2: Game Theory

1 William Poundstone, *Prisoner's Dilemma* (New York: Doubleday, 1992), 61–2; James D. Morrow, *Game Theory for Political Scientists* (Princeton: Princeton University Press, 1994), 89–91.
2 Bill James, *Major League Handbook, 1996* (Skokie, IL: STATS Publishing, 1995), 295–319. I have deducted at-bats taken by National League pitchers from this total.
3 John Maynard Smith, *Evolution and the Theory of Games* (Cambridge: Cambridge University Press, 1982), 10–27; Richard Dawkins, *The Selfish Gene*, 2nd ed. (Oxford: Oxford University Press, 1989), 66–87, 202–33, 282–7; Robert Axelrod, *The Evolution of Cooperation* (New York: Basic Books, 1984), 55–69.
4 Robert K. Adair, *The Physics of Baseball* (New York: HarperCollins, 1994), 96.
5 Willie Runquist, *Baseball by the Numbers* (Jefferson, NC: McFarland & Co., 1995), 34. Michael D. Akers, and Thomas E. Buttross, 'An Actuarial Analysis of the Production Function of Major League Baseball,' *Journal of Sport Behavior* 11 (1988), 99–112, found batting average to be the single most important variable in the 1980–4 data.
6 Stanley Coren, *The Left-Hander Syndrome: The Causes and Consequences of Left-Handedness* (New York: Vintage Books, 1992), 32. About 10 per cent of females are born left-handed.
7 Coren, *The Left-Hander Syndrome*, 212.
8 Dan Shaughnessy, 'Why Opportunity Beckons for Left-Handed Pitchers,' *Baseball Digest* 48 (July 1989), 29–33; Craig Adams, 'Why Left-Handed

Pitchers Are Prized in the Big Leagues,' *Baseball Digest* 49 (August 1990), 36–8.
9 Stephen R. Goldstein and Charlotte A. Young, '"Evolutionary" Stable Strategy of Handedness in Major League Baseball,' *Journal of Comparative Psychology* 110 (1996), 164–9; Thomas Flanagan, 'Game Theory and Professional Baseball,' *Journal of Sport Behavior*, 21 (1998), 121–38.

Chapter 3: Stalemate at Lubicon Lake

1 This section is a shortened and updated version of Thomas Flanagan, 'Aboriginal Land Claims in the Prairie Provinces,' in Ken Coates, ed., *Aboriginal Land Claims in Canada: A Regional Perspective* (Toronto: Copp Clark Pitman, 1992), 50–9. See also Flanagan, 'Some Factors Bearing on the Origins of the Lubicon Lake Dispute, 1899–1940,' *Alberta* 2 (1990), 47–62; Flanagan, 'The Lubicon Lake Dispute,' in Allan Tupper and Roger Gibbins, eds., *Government and Politics in Alberta* (Edmonton: University of Alberta Press, 1992), 269–303; and Flanagan, 'Adhesion to Canadian Indian Treaties and the Lubicon Lake Dispute,' *Canadian Journal of Law and Society* 7 (1992), 185–205. For a different view, see John Goddard, *The Last Stand of the Lubicon Cree* (Vancouver: Douglas & McIntyre, 1992).
2 *Ominayak v. Norcen*, 29 Alta. L.R. (2d) 152 (1984), at 157.
3 Boyce Richardson, 'The Lubicon of Northern Alberta,' in Boyce Richardson, ed., *Drumbeat: Anger and Renewal in Indian Country* (Toronto: Summerhill Press for the Assembly of First Nations, 1989), 246.
4 'Group Going to Top Court,' *Globe and Mail*, 11 May 1996.
5 Benoit Michel Clermont, 'An Analysis of the Legal Arguments Surrounding the Lubicon Land Claim' (MA thesis, University of Calgary, 1994), 29.
6 This section is a revised version of Flanagan, 'The Lubicon Lake Dispute,' 291–8.
7 This use of ordinal games follows Brams, *Rational Politics*, 7–23.
8 George Tsebelis, *Nested Games: Rational Choice in Comparative Politics* (Berkeley: University of California Press, 1990), 63.

Chapter 4: Models of Metrication

1 Minister of Industry, Trade, and Commerce, *White Paper on Metric Conversion in Canada* (Ottawa, January 1970).
2 Metric Commission, *Canada's Approach to Metric Conversion* (Ottawa, 1974).
3 N. Ganapathy, 'Metric Conversion,' in *The Canadian Encyclopedia*, 2nd ed. (Edmonton: Hurtig, 1988), 1346–7.

4 Daniel V. DeSimone, *A Metric America: A Decision Whose Time Has Come* (Washington: Department of Commerce, 1971).
5 David K. Lewis, *Convention: A Philosophical Study* (Cambridge, MA: Harvard University Press, 1962).
6 I here follow Robert Sugden, *The Economics of Rights, Co-operation and Welfare* (Oxford: Basil Blackwell, 1986), 34–54.
7 Witold Kula, *Measures and Men* (Princeton: Princeton University Press, 1986), 5.
8 Ibid., 18.
9 Henry Hamburger, *Games as Models of Social Phenomena* (New York: W.H. Freeman, 1979), ch. 7.
10 Iain McLean, *Public Choice: An Introduction* (Oxford, Basil Blackwell, 1987), 126.
11 Actually, it is the percentage of people *other than oneself*, but this distinction is not important with large numbers.
12 DeSimone, *A Metric America*, 26–7.
13 Kula, *Measures and Men*, 263–4.
14 *White Paper on Metric Conversion*, 5.
15 *The Weights and Measures Act*, R.S.C., 1985, c. W-6, s. 7(a).
16 Ibid., s. 10(1)(b).
17 Avinash Dixit and Barry J. Nalebuff, *Thinking Strategically: The Competitive Edge in Business, Politics, and Everyday Life* (New York: W.W. Norton, 1991), 243.
18 Alison Ramsey, 'Medical Alert: Childhood Diseases Are Back,' *Reader's Digest* (February 1996), 73–7.
19 This factor assumes that human beings are the only vectors of contagion. In fact, many diseases can be spread directly from other species of animals to humans. In such cases, it might still be rational to be vaccinated even if every other human being on earth were also vaccinated.
20 Arno Karlen, *Man and Microbes: Disease and Plagues in History and Modern Times* (New York: Simon & Schuster, 1995), 154–5. For a Canadian case study of the difficulties, see Michael Bliss, *Plague: A History of Smallpox in Montreal* (Toronto: HarperCollins, 1991), 207–15.

Chapter 5: How Many Are Too Many? The Size of Coalitions

1 Alexander H. Harcourt and Frans B.M. de Waal, eds., *Coalitions and Alliances in Humans and Other Animals* (Oxford: Oxford University Press, 1992), 3.
2 Mark O. Dickerson and Thomas Flanagan, *An Introduction to Government and Politics*, 4th ed. (Toronto: Nelson Canada, 1994), 11.

3 Henry Hamburger, *Games as Models of Social Phenomena* (New York: W.H. Freeman, 1979), 208–10.
4 William H. Riker, *The Theory of Political Coalitions* (New Haven: Yale University Press, 1962), 32–3.
5 Ibid., 89.
6 Ibid., 66–71.
7 Anthony Downs, *An Economic Theory of Democracy* (New York: Harper and Row, 1957), 11.
8 Riker, *Theory of Political Coalitions*, 32–3.
9 Ibid.
10 J.M. Beck, *Pendulum of Power: Canada's Federal Elections* (Scarborough: Prentice-Hall, 1968), 136–48; W.L. Morton, *The Progressive Party in Canada* (Toronto: University of Toronto Press, 1950), passim.
11 Beck, *Pendulum of Power*, 148–51; Morton, *Progressive Party*, passim.
12 Beck, *Pendulum of Power*, 243.
13 Conrad Black, *Duplessis* (Toronto: McClelland and Stewart, 1977), 404–9.
14 Recalculation of data in Maurice Pinard, *The Rise of a Third Party: A Study in Crisis Politics* (Montreal and Kingston: McGill-Queen's University Press, 1971), 30, Table 2.3.
15 John Sawatsky, *Mulroney: The Politics of Ambition* (Toronto: Macfarlane Walter and Ross, 1991), 472.
16 Patrick Martin, Allan Gregg, and George Perlin, *Contenders: The Tory Quest for Power* (Scarborough: Prentice-Hall, 1983), 200.
17 William Johnson, *A Canadian Myth: Quebec Between Canada and the Illusion of Utopia* (Montreal: Robert Davies, 1994), 197.
18 Tom Flanagan, *Waiting for the Wave: The Reform Party and Preston Manning* (Toronto: Stoddart, 1995), 39.
19 Robert Axelrod, *Conflict of Interest: A Theory of Divergent Goals with Applications to Politics* (Chicago: Markham, 1970), 167.
20 Ibid., 169.
21 Ibid., 170.
22 Axelrod, *Conflict of Interest*, 175–83; Michael Taylor and Michael Laver, 'Government Coalitions in Western Europe,' *European Journal of Political Research* 1 (1973), 222–6; Iain McLean, *Public Choice: An Introduction* (Oxford: Basil Blackwell, 1979), 121.
23 Abram De Swaan, *Coalition Theories and Cabinet Formations* (San Francisco: Jossey-Bass, 1973), 284–5. For the most recent game-theoretical approach to cabinet formation, see Michael Laver and Kenneth A. Shepsle, *Making and Breaking Governments: Cabinets and Legislatures in Parliamentary Democracies* (Cambridge: Cambridge University Press, 1996).

Chapter 6: Who's Got the Power? Amending the Canadian Constitution

1 Office of the Prime Minister, press release, November 27, 1995.
2 *The Constitution Act, 1982*, s. 41(e).
3 An earlier version of this chapter was published as 'Amending the Canadian Constitution: A Mathematical Analysis,' *Constitutional Forum* 7 (1996), 97–101. It contained an error in calculating the BPIs for the five-region veto. The error has now been corrected, so the statistics published here are not quite the same as in that article. For a similar approach with some differences of detail, see Andrew Heard and Tim Swartz, 'The Regional Veto Formula and Its Effects on Canada's Constitutional Amendment Process,' *Canadian Journal of Political Science* 30 (1997), 339–56.
4 *Canada Pension Plan*, R.S.C., 1985, c. 8, s. 114(4).
5 Steven J. Brams, *Rational Politics: Decisions, Games, and Strategy* (Boston: Harcourt Brace Jovanovich, 1985), 99.
6 D. Marc Kilgour, 'A Formal Analysis of the Amending Formula of Canada's *Constitution Act, 1982*,' *Canadian Journal of Political Science* 16 (1983), 773.
7 Ibid., 772.
8 See section 38(1) of the *Constitution Act, 1982*.
9 Heard and Swartz, 'The Regional Veto Formula,' 352.
10 Edward Greenspon and Anthony Wilson-Smith, *Double Vision: The Inside Story of the Liberals in Power* (Toronto: Doubleday Canada, 1996), 333–48.
11 When I had dinner with Stéphane Dion in Calgary a couple of days before he was appointed to the federal cabinet as minister of intergovernmental affairs, he pointed out to me the trap of legislating the five-region veto prematurely. He could not tell me then that he was about to enter the cabinet, so I did not fully appreciate his frustration until later.

Chapter 7: The 'Right Stuff': Choosing a Party Leader

1 For a simple explanation of the Condorcet winner, see J. Paul Johnston and Harvey E. Pasis, *Representation and Electoral Systems: Canadian Perspectives* (Scarborough: Prentice-Hall, 1990), 387–96.
2 *National Party Conventions, 1831–1972* (Washington: Congressional Quarterly, 1976), 8, 10.
3 John W. Lederle, 'The Liberal Convention of 1919 and the Selection of Mackenzie King,' *Dalhousie Review* 27 (1947), 90.
4 For a more detailed critique, see Peter C. Fishburn and Steven J. Brams, 'Paradoxes of Preferential Voting,' in Johnston and Pasis, *Representation and Electoral Systems*, 423–31.

5 On the campaign and convention, see Patrick Martin, Allan Gregg, and George Perlin: *Contenders: The Tory Quest for Power* (Scarborough: Prentice-Hall, 1983).
6 Terence J. Levesque, 'On the Outcome of the 1983 Conservative Leadership Convention: How They Shot Themselves in the Other Foot,' *Canadian Journal of Political Science* 16 (1983), 781.
7 My presentation of the arithmetic differs from that of Levesque, 'On the Outcome,' 781–3, but amounts to the same thing.
8 George Perlin, 'Did the Best Candidate Win? A Comment on Levesque's Analysis,' *Canadian Journal of Political Science* 16 (1983), 791–2.
9 Ibid., 793.
10 Keith Archer, 'Leadership Selection in the NDP,' in Herman Bakvis, ed., *Canadian Political Parties: Leaders, Candidates and Organization* (Toronto: Dundurn, 1991; vol. 13 of the research studies, Royal Commission on Electoral Reform and Party Financing), 20.
11 Ibid., 26. One thousand and fifty-five respondents gave a first choice; 930 carried it all the way down to seventh place.
12 Alan Whitehorn, 'How the NDP Chooses,' *Canadian Forum*, September 1995, 19. For more detail, see Keith Archer and Alan Whitehorn, *Political Activists: The NDP in Convention* (Toronto: Oxford University Press, 1997), 233–58.
13 David Roberts, 'Nystrom Well-Placed to Be NDP Leader,' *Globe and Mail*, 12 October 1995.
14 Telephone conversation with the author, April 1996. See also Archer and Whitehorn, *Political Activists*, 258, n. 41.
15 Hugh Winsor, 'Second Fiddle Leads NDP,' *Globe and Mail*, 16 October 1995.
16 Hugh Winsor, 'Feisty McDonough Raises Solidarity Issue,' *Globe and Mail*, 13 October 1996; Hugh Winsor, 'Tight Race Emerges for NDP Leadership,' *Globe and Mail*, 14 October 1996.
17 Steven J. Brams, *Rational Politics: Decisions, Games, and Strategy* (Boston: Harcourt Brace Jovanovich, 1985), 213.
18 Lawrence LeDuc, 'Party Decision-Making: Some Empirical Observations on the Leadership Selection Process,' *Canadian Journal of Political Science* 4 (1971), 100–3; Robert Krause and Lawrence LeDuc, 'Voting Behaviour and Electoral Strategies in the Progressive Conservative Leadership Convention of 1976,' *Canadian Journal of Political Science* 12 (1979), 101–6; Archer, 'Leadership Selection in the NDP,' 20–5.
19 Winsor, 'Second Fiddle Leads NDP.'
20 Tom Flanagan, *Waiting for the Wave: The Reform Party and Preston Manning* (Toronto: Stoddart, 1995), 141.

21 William Cross, 'Direct Election of Provincial Party Leaders in Canada, 1985–1995: The End of the Leadership Convention?' *Canadian Journal of Political Science* 29 (1996), 295–315.

Chapter 8: The Staying Power of the Status Quo

1 A somewhat different version of this chapter appeared as Thomas Flanagan, 'The Staying Power of the Legislative Status Quo: Collective Choice in Canada's Parliament after *Morgentaler*,' *Canadian Journal of Political Science* 30 (1997), 31–53.
2 *Morgentaler, Smoling and Scott v. the Queen*, [1988] 1 S.C.R. 30, reprinted in Research Unit for Socio-Legal Studies, University of Calgary, *Leading Constitutional Decisions of the Supreme Court of Canada*, 17.
3 Ibid., 13.
4 I rely on Janine Brodie, Shelley A.M. Gavigan, and Jane Jenson, *The Politics of Abortion* (Toronto: Oxford University Press, 1992); F.L. Morton, *Morgentaler v. Borowski: Abortion, the Charter, and the Courts* (Toronto: McClelland and Stewart, 1992); and Robert M. Campbell and Leslie A. Pal, 'Courts, Politics, and Morality: Canada's Abortion Saga,' in *The Real Worlds of Canadian Politics: Cases in Process and Policy*, 2nd ed. (Peterborough: Broadview Press, 1991).
5 Iain McLean, *Public Choice: An Introduction* (Oxford: Basil Blackwell, 1987), 25–7.
6 Note that these preference orderings are purely stipulative. In the real world, there could be conservatives whose second preference is f rather than l, and so on.
7 Steven J. Brams, *Rational Politics: Decisions, Games, and Strategy* (Boston: Harcourt Brace Jovanovich, 1985), 210.
8 William H. Riker, *Liberalism against Populism: A Confrontation between the Theory of Democracy and the Theory of Social Choice* (Prospect Heights, IL: Waveland Press, 1988/1982), 69–73.
9 Riker, *Liberalism against Populism*, 181–95; Riker, 'Implications from the Disequilibrium of Majority Rule for the Study of Institutions,' in Peter C. Ordeshook and Kenneth A. Shepsle, eds., *Political Equilibrium* (Boston: Kluwer-Nijhoff Publishing, 1982), 3–24; Peter C. Ordeshook, 'Comment on Riker,' ibid., 25–31; Fiorina and Kenneth A. Shepsle, 'Equilibrium, Disequilibrium, and the General Possibility of a Science of Politics,' ibid., 49–64; Jones, *Reconceiving Decision-Making in Democratic Politics*.
10 Brodie et al., *Politics of Abortion*, 67–8.

11 Graham Fraser, 'Tories Seek to Alter Debate Rules,' *Globe and Mail*, 21 May 1988.
12 William H. Riker, *The Art of Political Manipulation* (New Haven: Yale University Press, 1986), ix.
13 Graham Fraser, 'Opposition Kills Free Vote on Abortion,' *Globe and Mail*, 19 May 1988; *House of Commons Debates*, 24 May 1988, 15697ff.
14 *House of Commons Journals*, 28 July 1988, 3296–7; *House of Commons Debates*, 26 July 1988, 17964–46.
15 Details from *House of Commons Journals*, 28 July 1988, 3296–302.
16 Brodie et al., *Politics of Abortion*, 87.
17 In the figure, the few MPs who abstained on certain votes have been put into the category where they have the closest fit.
18 Brams, *Rational Politics*, 28.
19 Graham Fraser, 'MPs Vote Down Abortion Resolution,' *Globe and Mail*, 29 July 1988. See Link Byfield, 'A Pro-Life Coup in Ottawa,' *Alberta Report*, 8 August 1988, 38, for a similar view.
20 Brodie et al., *Politics of Abortion*, 61.
21 Gallup poll, 29 September 1988.
22 Sydney Sharpe, *The Gilded Ghetto: Women and Political Power in Canada* (Toronto: HarperCollins, 1994), 111–27.
23 Campbell and Pal, *Real Worlds*, 213.
24 Ibid., 43.
25 Brodie et al., *Politics of Abortion*, 99, 168 (n. 81).
26 *House of Commons Debates*, 28 November 1989, 6342.
27 McLean, *Public Choice*, 133–4.
28 *House of Commons Journals*, 28 November 1989, 897, and 29 May 1990, 1770–1.
29 Frith in *Senate Debates*, 7 June 1990, 1866; Fairbairn in ibid., 29 January 1991, 5209–13; Kirby in ibid., 29 January 1991, 5217–22. Neiman did not speak but seconded a pro-choice amendment introduced by Kirby, ibid., 5222.
30 Bélisle in ibid., 6 June 1990; Bosa in ibid., 24 January 1991, 5181–2; Ray Perrault in ibid., 30 January 1991, 5259–60; Haidasz in ibid., 26 June 1990, 2149–53, and 29 January 1991, 5233–43. J.-P. Guay also expressed opposition on pro-life grounds (ibid., 6 June 1990), but was no longer in the Senate when the bill was voted upon. The other names will be found in the votes on the Haidasz amendments as recorded in the *Senate Journals*, 31 January 1991, 2230–7.
31 *Senate Debates*, 13 June 1990, 1932–3.
32 Ibid., 5 June 1990, 1828.
33 Ibid., 29 January 1991, 5224. See also the remarks of Solange Chaput-Rolland, ibid., 29 January 1991, 5236–7.

34 Geoffrey York, 'Senators Kill Abortion Bill with Tied Vote,' *Globe and Mail*, 1 February 1991. In June 1996, the Liberals' Pearson Airport bill was also defeated on a tie vote in the Senate.
35 *The Constitution Act, 1867*, s. 49; *Beauchesne's Rules and Forms of the House of Commons of Canada*, 5th ed. (Toronto: Carswell, 1978), s. 226.
36 W.H. McConnell, *Commentary on the British North America Act* (Toronto: Macmillan of Canada, 1977), 79.
37 *The Constitution Act, 1867*, ss. 34, 36.
38 F.A. Kunz, *The Modern Senate of Canada, 1925–1963: A Reappraisal* (Toronto: University of Toronto Press, 1965), 168.
39 *Senate Debates*, 30 January 1991, 5260.

Chapter 9: Invasion from the Right: The Reform Party in the 1993 Election

1 Britain, Canada, and the United States use the first-past-the-post system. Australia's alternative ballot is similar in practice, except that it has promoted a two-party coalition (Liberal/National) on the right. New Zealand adopted a form of proportional representation in 1993, but it was not actually used until the election of 1996.
2 Patrick Donleavy, *Democracy, Bureaucracy and Public Choice* (New York: Prentice-Hall, 1991), 132–5.
3 From this point forward, the chapter draws heavily on Thomas Flanagan, 'Invasion from the Right: The Reform Party in the 1993 Canadian Election,' *Papers in Political Economy* (Political Economy Research Group, University of Western Ontario), no. 42, 1994.
4 Réjean Landry, 'Incentives Created by the Institutions of Representative Democracy,' in Herman Bakvis, ed., *Representation, Integration and Political Parties in Canada* (Toronto: Dundurn, 1991; vol. 14 of the Research Studies, Royal Commission on Electoral Reform and Party Financing), 446–8; Steven J. Brams, *Rational Politics: Decisions, Games, and Strategy* (Boston: Harcourt Brace Jovanovich, 1985), 32–6.
5 Anthony Downs, *An Economic Theory of Democracy* (New York: Harper and Row, 1957), 131.
6 Keith Archer and Faron Ellis, 'Opinion Structure of Reform Party Activists,' Midwest Political Science Association, Chicago, April 1993, Table 4.
7 Re-analysis of Archer and Ellis data (ibid.) by Michael Wagner.
8 Thomas Flanagan and Faron Ellis, 'A Comparative Profile of the Reform Party of Canada.' Paper presented to the Canadian Political Science Association, Charlottetown, June 1992, 8.
9 E.C. Manning, *Political Realignment: A Challenge to Thoughtful Canadians* (Toronto: McClelland and Stewart, 1967), 56–70; Thomas Flanagan and

Martha Lee, 'From Social Credit to Social Conservatism: The Evolution of an Ideology,' *Prairie Forum* 16 (1991), 205–23.
10 Preston Manning, 'Choosing a Political Vehicle to Represent the West,' in Ted Byfield, ed., *Act of Faith* (Vancouver: *BC Report* Magazine, 1991), 172.
11 Preston Manning, 'Building the Reform Team: The Hockey Analogy,' in Byfield, *Act of Faith*; see also George Koch, 'Looking Leftward: Manning's Search for New Members Angers the Party's Right Wing,' *Alberta Report*, 5 February 1990, in *Act of Faith*, 89–90.
12 Preston Manning, 'An Open Letter to the Members of Local 222, Canadian Auto Workers,' 6 August 1993.
13 Sean Durkan, 'PM Begins Race in Lead,' *Calgary Sun*, 9 September 1993; 'Grits, Tories Start Race in Virtual Tie,' *Calgary Herald*, 11 September 1993.
14 Edward Greenspon and Jeff Sallot, 'How Campbell Self-destructed,' *Globe and Mail*, 27 October 1993.
15 Susan Delacourt, Murray Campbell, and Edward Greenspon, 'Liberal Hopes on the Rise,' *Globe and Mail*, 21 September 1993.
16 Jeff Sallot and Hugh Winsor, 'PM Won't Touch Key Issue,' *Globe and Mail*, 24 September 1993.
17 Joe Sornberger, 'PM Promises More Info on Cuts,' *Calgary Herald*, 25 September 1993.
18 David Steinhart, 'Deficit Plan Unveiled,' *Calgary Herald*, 28 September 1993.
19 Edward Greenspon, 'Tories' Supposed Strong Point Turning Out to Be Anything But,' *Globe and Mail*, 29 September 1993.
20 Warren Caragata, 'Debate: Insults Outweigh Ideas,' *Calgary Herald*, 5 October 1993.
21 Ross Howard, 'Tories Rush to Attack with New Ads,' *Globe and Mail*, 6 October 1993.
22 Sean Durkan and Bill Kaufmann, 'Heat's on Reform' *Calgary Sun*, 10 October 1993.
23 Tim Naumetz, 'Cuts Made on the Run,' *Calgary Sun*, 29 September 1993.
24 'The Voters Speak,' *Globe and Mail*, 25 October 1993.
25 Edward Greenspon, Ross Howard, and Susan Delacourt, 'Tories Try to Recover from Goof,' *Globe and Mail*, 16 October 1993; Julian Beltrame, 'Kim Targets Tories,' *Calgary Herald*, 17 October 1993.
26 Reform Party news release, 'Manning Says: "Let the People Speak!"' 8 September 1993.
27 Norm Ovenden, 'Manning Outlines Deficit Plan,' *Calgary Herald*, 21 September 1993.
28 'The Only Deficit Plan We've Seen,' *Globe and Mail*, 23 September 1993.
29 Hugh Winsor, 'Liberals Near Majority, Globe Poll Finds,' *Globe and Mail*, 16 October 1993.

30 Environics tracking poll, 14 October 1993, three-day average.
31 Hugh Winsor, 'Liberals Near Majority, Globe Poll Finds,' *Globe and Mail*, 16 October 1993.
32 Reform Party news release, 'Manning says: "Liberals Don't Deserve a Majority Government; Minority Parliament is Best for Canada,"' Cambridge/Sarnia, 12 October 1993.
33 Reform Party news release, 'Manning Offers Canada New Federalism,' Toronto, 1 October 1993.
34 Reform percentages for 1993 were calculated from returns printed in the *Globe and Mail*, 27 October 1993. Other data were taken from Monroe Eagles, James P. Bickerton, Alain-G. Gagnon, and Patrick J. Smith, *The Almanac of Canadian Politics* (Peterborough: Broadview Press, 1991).
35 Harold Clarke, The Dynamics of Support for New Parties and National Party Systems in Contemporary Democracies: The Case of Canada. Funded by the National Science Foundation. Data courtesy of Harold Clarke.
36 On a goodness-of-fit test, chi square = 48.6, $df = 2$, $p < 0.0000$.
37 On a goodness-of-fit test, chi square = 25.9, $df = 2$, $p < 0.0000$.
38 Neil Nevitte et al., 'Electoral Discontinuity: The 1993 Canadian Federal Election,' *International Social Science Journal* 146 (December 1995), 583–99.
39 Ibid., Table 3, 589.
40 The direction was reversed only on cuts to health care, and the difference was small: 2.1 per cent.
41 Arend Lijphart, *Democracies: Patterns of Majoritarian and Consensus Government in Twenty-One Countries* (New Haven: Yale University Press, 1984), 127–49.
42 Riker, *Liberalism against Populism*, 197–232.
43 Preston Manning, 'Choosing a Political Vehicle to Represent the West,' in Byfield, *Act of Faith*, 171.
44 Manning, *The New Canada* , 6–7.
45 Reform Party of Canada, Statement of Principles, no. 14, in the *Blue Sheet* [1993], 2.
46 Preston Manning, tape recording, 'Countdown to Victory,' 2 December 1993.
47 Reform Party, *Principles and Policies*, 1990, 7; Paul Bunner and Peter MacDonald, 'A National Agenda: Manning Calls Quebec's Bluff,' *Alberta Report*, 6 November 1989, in Byfield, *Act of Faith*, 76–8.
48 Nevitte et al., 'Electoral Discontinuity,' Table 3, 589.
49 Joan Bryden, 'Reform's "Ship" Dead in the Water,' *Ottawa Citizen*, 8 February 1993. Manning used the same analogy in a taped 'Fireside Chat' to all candidates, 12 February 1993.

Chapter 10: What Have We Learned?

1 Anatol Rapoport, Melvin J. Guyer, and David G. Gordon, *The 2 X 2 Game* (Ann Arbor: University of Michigan Press, 1976).
2 Donald P. Green and Ian Shapiro, *Pathologies of Rational Choice Theory: A Critique of Applications in Political Science* (New Haven: Yale University Press, 1994).
3 John Stuart Mill, *Principles of Political Economy and Chapters on Socialism* (Oxford: Oxford University Press, 1994), 349.
4 Iain McLean, *Public Choice: An Introduction* (Oxford: Basil Blackwell, 1987), 126–32.

Index

Abortion 120–39
Altruism 10
Archer, Keith 112–13, 117
Aristotle 14
Assurance game 61–2, 169
Axelrod, Robert 88, 91, 140

Backwards induction 23
Banzhaf Power Index (BPI) 94–8, 101, 165
Barrett, Dave 112–13, 117
Baseball 11, 30–6
Beetz, Jean 120
Behavioural research 16–17
Bélisle, Rhéal 136–7
Bloc Populaire 84, 92
Bloc Québécois 87, 150–3, 162
Bonchek, Mark S. 4
Borden, Robert 79, 83
Bosa, Peter 136
Bosley, John 126–7
Bouchard, Benoît 86
Bouchard, Lucien 86, 150
Brams, Steven J. 95, 145–6
Branches 22
Brennan, Geoffrey 10
Broadbent, Ed 112, 118, 154

Brodie, Janine 130–1
Buchanan, James M. 10

Campbell, Kim 150–2
Canada Pension Plan 94
Caouette, Réal 85
Cardinal utility 7–8
Carney, Pat 131, 135
Charbonneau, Guy 137
Charest, Jean 151
Chicken game 71, 133, 166, 169
Chrétien, Jean 93, 103, 144, 151
Christian Heritage Party 148
Clark, Joe 86, 110–12, 117–18
Clarke, Harold 17, 156–8
Coalition government 89–92
Coalitions 24, 74–92, 169–70
Collins, Mary 126–7, 131
Comparability 6
Comte, Auguste 18–19
Condorcet, Marquis de 105
Condorcet (extension) 122–4, 166; (winner) 105–7, 109–18, 121–4
Confederation of Regions Party 148
Constitutional amendment (7/50 formula) 97–9, 102
Continuous choice 140

Conventions 58–60
Conversion, metric 56
Co-operative Commonwealth Federation (CCF) 80, 84, 119, 147, 160
Coordination game 58–60, 169
Cotret, Robert de 151
Crerar, T.A. 83
Crombie, David 43
Crosbie, John 110–12, 117
Cycles 107, 121–4

Daishowa 45–6
Deadlock game 39, 49, 167–8
De Jong, Simon 112
Democratic Party 108
Diefenbaker, John 84–6, 91–2, 109, 119
Discrete choice 140
Distance principle 88–91
Distinct society 15, 103–4
Divide the Dollar game 74–5
Doer, Gary 114, 117
Dominant strategy 21–2, 38
Doody, William 136
Downs, Anthony 4, 10, 78, 140–4, 147, 159
Duplessis, Maurice 85

Ecological fallacy 156
Equilibrium (unstable) 64; (structure-induced) 123–4, 129
European Union 95
Exhaustiveness 6
Expected value 8–9
Extensive form 22–3

Fairbairn, Joyce 136
Fiorina, Morris 124
Five-region veto 93–4, 99–104

Forsyth, Gregory 42
Franks, C.E.S. 15–16
Free Trade Agreement 132, 144
Frith, Royce 136
Fulton, E. Davie 43

Game theory (history of) vii–viii; (essential ideas of) 20–36
Game tree 22–3
Getty, Don 44–5
Glenbow Alberta Institute 44
Graham, Bernard Alasdair 136
Grand coalition 74–5
Green, Donald P. 168
GST 135, 144
Gzowski, Peter 151

Haidasz, Stanley 136
Harris, Mike 119
Heresthetic 125, 133
Homo economicus 3

Imputation 75
Indian Association of Alberta 41
Individualism, methodological 4–5
Information 11, 21, 146–7
Interdependence 13
Iterated deletion of dominated strategies 25–6

James, Ken 126–7, 131
Johnson, Daniel 104
Johnson, Janis 136
Johnston, Richard 158
Jones, Bryan D. 124

Kelly, William 136
Kilgour, D. Marc 97–8
King, William Lyon Mackenzie 83, 92
Kirby, Michael 136

Index 189

Klein, Ralph 119
Kula, Witold 66

Landry, Réjean 145-6
Langdon, Steven 112-13, 117
Laurier, Wilfrid 83
League of Nations 78
Lefebvre, Thomas-Henri 136
Lennarson, Fred 44
Levesque, Terence 110-11
Lewis, Doug 124-6
Liberal Party 78-88, 92, 108, 148-63
Loon River Cree 46
Lougheed, Peter 41
Lubicon Lake conflict 40-54, 167-8

MacDonald, Flora 131
Macdonald, John A. 138
Macdonald, John Michael 136
Majority coalition 74-5
Malone, Brian 44, 46
Manning, Ernest 148
Manning, Preston 148, 150-2, 156, 159-62
Masse, Marcel 86
Mazankowski, Don 151
McDonough, Alexa 114-19
McDougall, Barbara 131
McKenna, Frank 170
McKnight, Bill 44
McLaughlin, Audrey 112-13, 118-19
Median voter 128, 138, 140-5
Meighen, Arthur 83-4
Metrication 55-70, 168-9
Microfoundations 5, 71
Mill, John Stuart 3, 169
Minimax. *See* Mixed-strategies, solution in
Minimum blocking coalition 97
Minimum winning coalition 75-6, 89, 95-6, 99-100; (compact) 90; (connected) 89
Minority coalition 74-5
Mitges, Gus 126-9, 131, 133
Mixed-strategies, solution in 27-30, 33-6
Molgat, Gildas 136
Morgenstern, Oskar vii
Morgentaler case 120-1, 138
Mulroney, Brian 86-7, 91, 110-12, 118, 120, 124, 131-5, 151, 166

N-person games 24, 74-6
Nash, John vii
Nash equilibrium 38-9, 48, 50, 59, 60, 62, 64, 71
Neiman, Joan 136
Neumann, John von vii
New Democratic Party (NDP) 80, 112-19, 147-9, 154-8, 160, 166
New Zealand 144
Nodes 22
Nominating conventions 108-19
Norcen 45
Normal form 21
Null coalition 74-5
Nunziata, John 78
Nystrom, Lorne 114-18

O'Reilly, James 42, 45
Olerud, John 11
Olson, Mancur 4
Ominayak, Bernard 42-5
Ordinal games 26, 47-54
Ordinal utility 5-6

Paradox of voting 121-4
Pareto optimum 38-9, 62
Parti Québécois 91
Party discipline 15-16, 78, 167

Payoff 20
Pearson, Lester 84
Perlin, George 111–12, 117
Perot, Ross 142
Perrault, Ray 136–7
Petro-Canada 45
Plato 14
Populism 160–1
Precommitment 133–5
Prisoner's Dilemma game 37–9, 169
Progressive Conservative Party 79–88, 91–2, 109–12, 148–63
Progressive Party 80–1, 160

Rae, Bob 114, 116
Rational choice 3–19
Rationality 7
Reform Party 80, 87, 92, 147–63, 170
Republican Party 108
Riker, William H. vii, 76–8, 80, 82–3, 88, 124–5, 169
Robinson, Svend 114–17
Romanow, Roy 114

Schelling, Thomas 62, 68
Schelling curves 62–4, 70–3
Self-interest 9–10, 164
Shapiro, Ian 168
Shepsle, Kenneth A. 4, 124
Simon, Herbert 13
Size principle 76–8; (empirical tests of) 79–83, 88
Social Credit Party 80, 85–8, 148, 160, 170
Solution concept 20
Solution set 75
Sparrow, Bobbie 126

Spatial models 89–90, 140–7, 162–3, 170
Speaker (of the Senate) 136–8
Spivak, Mira 135–6
St Laurent, Louis 85, 109
Stanfield, Robert 86, 118
Strategic form 21
Strategic voting 115–18, 154
Strategy 20
Supreme Court of Canada 120–1, 138–9

Tassé, Roger 43–4
Taylor, Charles 14–15
Teresa, Mother 9
Thatcher, Margaret 144
Thériault, L. Norbert 136
Tipping 67–8
Transitivity 6–7, 107
Trap 62
Treaty Eight 40, 42–3
Trudeau, Pierre 43, 55
Turner, John 132
Two-person game 36

Union government 77, 79, 83
Union Nationale 85, 91
United Nations 78, 95
Utility function 5–6

Vaccination 70–3
Variable-sum games 24, 37–8

Weighted voting 75–6, 94–7
Wilson, Bertha 120
Winsor, Hugh 116
Woodland Cree 46, 53

Zero-sum games 24, 30, 37